Education and Training in Indoor Air Sciences

NATO Science Series

A Series presenting the results of activities sponsored by the NATO Science Committee. The Series is published by IOS Press and Kluwer Academic Publishers, in conjunction with the NATO Scientific Affairs Division.

A. Life Sciences	IOS Press
B. Physics	Kluwer Academic Publishers
C. Mathematical and Physical Sciences	Kluwer Academic Publishers
D. Behavioural and Social Sciences	Kluwer Academic Publishers
E. Applied Sciences	Kluwer Academic Publishers
F. Computer and Systems Sciences	IOS Press

1. Disarmament Technologies	Kluwer Academic Publishers
2. Environmental Security	Kluwer Academic Publishers
3. High Technology	Kluwer Academic Publishers
4. Science and Technology Policy	IOS Press
5. Computer Networking	IOS Press

NATO-PCO-DATA BASE

The NATO Science Series continues the series of books published formerly in the NATO ASI Series. An electronic index to the NATO ASI Series provides full bibliographical references (with keywords and/or abstracts) to more than 50000 contributions from internatonal scientists published in all sections of the NATO ASI Series.
Access to the NATO-PCO-DATA BASE is possible via CD-ROM "NATO-PCO-DATA BASE" with user-friendly retrieval software in English, French and German (WTV GmbH and DATAWARE Technologies Inc. 1989).

The CD-ROM of the NATO ASI Series can be ordered from: PCO, Overijse, Belgium

2. Environmental Security – Vol. 60

Education and Training in Indoor Air Sciences

edited by

Nadia Boschi

Department of Building Construction,
College of Architecture and Urban Studies,
Virginia Polytechnic Institute & State University,
Falls Church (VA), U.S.A.

Kluwer Academic Publishers

Dordrecht / Boston / London

Published in cooperation with NATO Scientific and Environmental Affairs Division

Proceedings of the NATO Advanced Research Workshop on
Education and Training in Indoor Air Sciences
Budapest, Hungary
November 14–18, 1998

A C.I.P. Catalogue record for this book is available from the Library of Congress.

ISBN 0-7923-5910-0 (HB)
ISBN 0-7923-5911-9 (PB)

Published by Kluwer Academic Publishers,
P.O. Box 17, 3300 AA Dordrecht, The Netherlands.

Sold and distributed in North, Central and South America
by Kluwer Academic Publishers,
101 Philip Drive, Norwell, MA 02061, U.S.A.

In all other countries, sold and distributed
by Kluwer Academic Publishers,
P.O. Box 322, 3300 AH Dordrecht, The Netherlands.

Printed on acid-free paper

TABLE OF CONTENTS

vi

CONTRIBUTORS

JUDIT BÁCSKAI, National Institute of Environmental Health, Jozsef Fodor National Public Health Centre, Hungary

WIM BAKENS, International Council for Building Research Studies and Documentation (CIB), General Secretariat, The Netherlands

ZSOLT BAKO, Department of Building Services Engineering, Technical University of Budapest, Hungary

LÁSZLO BÁNHIDI, Department of Building Services Engineering, Technical University of Budapest, Hungary

GABRIELLA M. BELLI, College of Human Resources and Education, Virginia Polytechnic Institute and State University, USA

VLADIMÍR BENCKO, Institute of Hygiene and Epidemiology, First Faculty of Medicine, Charles University of Prague, Czech Republic

PHILOMENA M. BLUYSSEN, Department of Indoor Environment, Building Physics and System, TNO Building and Construction Research, The Netherlands

NADIA BOSCHI, Department of Building Construction, Virginia Polytechnic Institute and State University, USA

HOWARD S. BRIGHTMAN, School of Public Health, Harvard University, USA

ANNA CHARKOWSKA, Institute of Heating and Ventilation, Warsaw University of Technology, Poland

GEO CLAUSEN, Centre for Indoor Environment and Energy, Technical University of Denmark, Denmark

DPHIL CPSYCHOL, Building Research Establishment Ltd., United Kingdom

ILDIKO FARKAS, National Institute of Environmental Health, Jozsef Fodor National Public Health Centre, Hungary

ELISSA FELDMAN, Indoor Environments Division, U.S. Environmental Protection Agency, USA

MELVIN W. FIRST, School of Public Health, Harvard University, USA

JOZSEF FODOR, National Public Health Centre, National Institute of Environment Health, Hungary

GIOVANNA GUARNERIO, DIPRA, Polytechnic of Turin, Italy

FARIBORZ HAGHIGHAT, Department of Building, Civil and Environment Engineering, Concordia University, Canada

IVANA HOLCÁTOVÁ, Institute of Hygiene and Epidemiology, First Faclty of Medicine, Charles University of Prague, Czech Republic

PENTTI J. KALLIOKOSKI, Department of Environmental Sciences, University of Kuopio, Finland

ILDIKO KATONA, City Public Health Institute, Százhalombatta, Hungary

SØREN K. KJÆRGAARD, Department of Environmental and Occupational Medicine, Aarhus University, Denmark

LASZLO KŐRÖSI, Outpatient Clinic, Százhalombatta, Hungary

HAL LEVIN, Building Ecology Research Group, USA

MARCO MARONI, University of Milano, and WHO/ International Centre for Pesticide Safety, Italy

STEFAN MAZIARKA, Department of Environmental Hygiene, National Institute of Hygiene, Poland

ILDIKO MOCSY, Institute of Public Health Cluj, Romania

LARS MØLHAVE, Department of Occupational and Environmental Medicine, Aarhus University, Denmark

LIDIA MORAWSKA, School of Physical Sciences, Queensland University of Technology, Australia

GIACOMO MUZI, Institute of Occupational Medicine and Occupational and Environmental Toxicology, University of Perugia, Italy

ANNA PÁLDY, National Institute of Environmental Health, Jozsef Fodor National Public Health Centre, Hungary

KATALIN PAPP, City Public Health Institute, Gödöllő, Hungary

RACHELE PAVESI, DIPRA, Polytechnic of Turin, Italy

DUSAN PETRAS, Department of Building Services, Slovak Technical University, Slovak Republic

ALALN PÍNTÉR, National Institute of Environmental Health, Jozsef Fodor National Public Health Centre, Hungary

GARY J. RAW, Building Research Establishment Ltd., United Kingdom

PETER RUDNAI, National Institute of Environmental Health, Jozsef Fodor National Public Health Centre, Hungary

INGRID SENITKOVA, Civil Engineering Institute, Technical University Kosice, Slovak Republic

KATARINA SLOTOVA, State Health Institute, Slovak Pepublic

JOHN D. SPENGLER, School of Public Health, Harvard University, USA

JAN SUNDELL, International Centre for Indoor Environment and Energy, Technical University of Denmark, Denmark, and National Institute of Public Health, Sweden

FERENC SZÍJJÁRTÓ, General Practitioner, Szgetúifalu, Hungary

MARIA TCHOUTCHKOVA, Urban Environment and Health Section, National Centre of Hygiene, Medical Ecology and Nutrition, Bulgaria

DAVID P. WYON, Centre for Indoor Environment and Energy, Technical University of Denmark, Denmark

MAGYAR ZOLTAN, General Practitioner, Hungary

PREFACE

This book deals with education in Indoor Air Sciences (IAS). The intent of the book is to present the results of a project conducted under the auspices of the Scientific and Environmental Affairs Division of the North Atlantic Treaty Organization (NATO). The audience for this subject continues to grow, as the dimension of adverse health effects due to indoor environment exposures becomes more and more apparent. The growing impact of this problem on people and properties has shown the limitations and lack of comprehensiveness that current educational paradigms for building and health professionals have in addressing the building-occupant relationship.

The issues related to education in IAS are many, and the large number of disciplines involved in it includes architects, clinicians, engineers, physicists, psychologists and policy makers. A research workshop, held in Budapest in November 1998, constitutes a key component of the project itself. The workshop convened 28 participants from 14 different countries. The purpose of the workshop was to discuss examples of existing educational programs which aim to bridge the existing gap between health and building sciences and define a common core curriculum for education in IAS.

The structure of this book reflects the complexity of education in IAS. The book is organized in 9 parts. Part 1 constitutes a brief introduction to the issue of education in IAS. Part 2 provides an historical perspective on framing the field in IAS. Part 3 discusses key current trends that are leading education in IAS. Then Part 4 presents a representative sample of existing educational programs currently available in the field of IAS at institutions of higher education, in government agencies and other professional settings in North- American, Australian, European and Scandinavian countries. Part 5 presents some programs that propose the use of research and problem based approaches to provide education in IAS. New approaches and emerging issues in IAS are discussed in Part 6. Part 7 identifies the different players involved in IAS education and compares educational needs in different cultural and geographical contexts. Part 8 describes the methodology used to conduct the workshop. Finally, Part 9 presents a summary and the conclusions obtained in conducting this project.

Overall, the project's conclusion was that a core curriculum for IAS is identifiable. How detailed each topic has to be developed and how much emphasis each would have relative to other topics has to be determined by the audience, profession or objective of a course. Obstacles to its implementation exist at the policy, academic and professional level and the strategies identified need to be considered in short and long-term feasible scenarios.

Finally, it should be noted that the conclusions and recommendations of this study are based on the knowledge and expertise available at this time. A great need for research and further understanding of educational needs in IAS remains. It is my hope that this book will be a call to action. Many of the adverse health effects caused by indoor environments can be prevented and education has been proven to be an effective tool for prevention. What is needed is a plan that aims to raise public awareness and ensure appropriate education and training for professionals involved in the health and building sector.

NB

ACKNOWLEDGMENTS

I wishes to acknowledge the North Atlantic Treaty Organization (NATO) Scientific and Environmental Affairs Division for the support given to the Advanced Research Workshop titled *"Education and Training in Indoor Air Sciences"*. The workshop was held in Budapest (Hungary) in November 14-18, 1998.

In addition, I wish to thank and express my gratitude for the support and assistance very helpful and generous scientists. Dr. Marco Maroni, University of Milan (Italy), for his guidance, encouragement and support in my attempt in seeking international and interdisciplinary education that integrates health and building sciences. Dr. László Bánhidi, Technical University of Budapest (Hungary), and Dr. Peter Rudnai, National Institute of Public Health (Hungary), that co-directed the project and offered unselfish guidance in learning about working and living in Hungary. To the other members of the Workshop Organizing Committee for the indispensable advice provided: Dr. Lars Mølhave, Aarhus University (Denmark); Dr. Demetrios J. Moschandreas, Illinois Institute of Technology (US); and Dr. Olli Seppänen, Helsinki Technical University (Finland).

Further, a debt of gratitude is owed to all workshop participants without whose support I could not have conducted the project: Z. Bako, L. Banhidi, G. Belli, V. Bencko, P. Bluyssen, H. S. Brightman, G. Clausen, I. Farkas, E. Feldman, H. Haghigat, I. Holcatova, P. Kalliokoski, S. Kjærgaard, H. Levin, M. Maroni, S. Maziarka, I. Mocsy, L. Molhave, G. Muzi, A. Paldy, D. Petras, G. Raw, P. Rudnai, I. Senitkova, J. Sundell, M. Tchoutchkova and M. Zoltan.

Great appreciation to the friend and colleague Hal Levin for his support in preparing for the workshop, during the workshop, and, finally, for reviewing the final version of this book.

Lastly, I would like to special acknowledge the contributions of Ruangsak Sriwatthanah who provided administrative and organizational support with excellent skills and dedication.

PART I. INTRODUCTION

DEFINING AN EDUCATIONAL FRAMEWORK FOR INDOOR AIR SCIENCES EDUCATION

N. BOSCHI, Ph.D., Architect
Virginia Polytechnic Institute and State University
7054 Haycock Road, Falls Church, VA 22043-2311 USA

> *"The living conditions determine to a great extent the quality of life, which improvement is important for meeting the main needs of work, house, health service, education and recreation"*
> (Vancouver Declaration on Settlements, 1976)

Introduction

This work began with the hypothesis that buildings are designed and built to enhance our quality of life. Many buildings fail to perform adequately and cause illnesses and productivity losses to the occupants. Studies in the United States (US) and Europe have made the adverse health effects caused by indoor environmental exposure become more and more apparent. Also, while many buildings are constructed each year, approximately 90 % of the buildings that will exist in the developed countries during the first quarter of the next century have already been built. The growing impact of the indoor-pollution problem on people and property values - and its rise as the subject of litigation - have shown the limitations and the lack of comprehensiveness that current educational paradigms for building and health professionals have in addressing the building-occupant relationship. Thus, educational curricula for building and health professionals should include points of interface where health professionals know more about building systems, and building professionals understand more about human response.

From an educational perspective, current educational paradigms for design, construction and operation have significant limitations with regard to their effectiveness in addressing environmental health and well being of occupants. Likewise, educational paradigms for physicians, exposure experts, epidemiologists and other life sciences have significant limitations with regard to their effectiveness in addressing building systems [1]. Furthermore, the increasing difficulty for health and building professionals to control the large amount and disparate nature of information involved in delivering and operating buildings emphasizes the need for a reorganization of conventional curricula. Similar to what happened in medicine and law, a century ago, building professionals are now facing

3

a reality that is complex and requires a comprehensive understanding of buildings as well as availability of specialists in certain matters such as design, technology, construction management, and building epidemiology.

Although the concepts of sick and healthy buildings have helped the academic community to recognize the importance of integrating health and building sciences, they have also caused controversy, especially among those who are responsible for the financing, design, construction, and operation of buildings and their systems; those responsible for public health; and those who are responsible for medical treatment of occupants. Furthermore, integration of health and building sciences has proven to be a difficult educational and professional task because of the differences in the methods, technical languages, and cultural references.

Further, most educational programs now available that address environmental issues often do not adopt state-of-the-art procedures and, in most cases, do not include an overall evaluation of how the building's performance affects the health and well-being of the occupants or the building owners and managers. The protocols taught and employed focus on limited aspects of human response and exposures but seldom provide for comprehensive evaluation of system or economic performance.

In addition, educational programs, studies and investigations, even within one discipline, are taught and conducted using different educational paradigms, different diagnostic procedures and different sets of evaluation criteria leading to difficulties in comparing findings and reaching conclusions.

The results of the project presented in this report build on a concept of the *International School of Indoor Air Sciences*. This School constitutes an effort carried on by a worldwide net of universities to foster the development of a harmonized international culture on indoor air sciences (IAS). The School represents a unique opportunity to break cultural as well as geographical barriers, and offers an answer to the scientific and economic necessity to operate professionally on a scale that overcomes time and space barriers. The purpose of the School is achieved through cooperative development of educational programs [1].

Project' s Objectives

This publication, *Education and Training in Indoor Air Sciences,* provides a comprehensive review of the results obtained in conducting a project on education in IAS. The purpose of the project was to obtain an international perspective on existing educational programs that aim to bridge the gap between health and building sciences and define a framework for a common core curriculum for education in IAS. Specifically, the project had the following tasks:

1. Examine and characterize a representative sample of existing educational programs currently available in the field of IAS at institutions of higher education, in government agency and other professional settings;

2. Identify and discuss current trends, approaches and emerging issues that are characterizing education in IAS;

3. Identify different players involved in IAS education and compare educational needs in different cultural and geographical contexts;

4. Provide a forum for advanced level educators to define a conceptual framework for international curricula that could be integrated in existing university programs;

5. Identify possible strategies for establishing educational programs in IAS in different cultural and geographical contexts.

General approach

The project included an initial call for paper, a workshop, and a follow -up phase of analysis of the data gathered during the workshop. The workshop, held in Budapest in November 1998, constitutes the key component of the project itself. The purpose of the workshop was to convene educators in the field of architecture, engineering, medicine, physics, chemistry, psychology, and public health, along with policy makers to generate a core curriculum for education in IAS.

The workshop was planned around four major parts. The first part consisted of oral presentations to obtain an overview of existing educational programs, trends and emerging issues. The second part to create a forum for the participants to interact and reach consensus on important aspects of defining a core curriculum. The third part consisted of three (3) focus group sessions, to define: a) topics that should be in a common core curriculum for indoor air professionals; b) obstacles to implementing a basic core curriculum; c) strategies to overcome the obstacles. The fourth part was identification of action plans for the implementation of the core curriculum and recommendations for future developments.

The workshop started with the participants' oral presentations. Participants were familiar with the scope of the project and were asked to contribute on specific subject matters, although, the contributions were presented following a round-table format. Questions and discussions on the presentations were all deferred to the "forum session" held the following day.

The second part of the workshop constituted an essential step to bridge the communication gap between the participants. This phase was focused on creating a forum for the participants to interact and reach consensus on important aspects of defining a core curriculum. Participants were asked to discuss the need for developing a core curriculum for IAS, the rationale for developing a core curriculum, the audience to be addressed by the core curriculum as well as the criteria to define the curriculum. The different perspectives and issues raised by the participants constituted the stepping stone for the discussion in defining the workshop tasks.

6

The third part of the workshop was conducted using a qualitative research methodology. Qualitative data gathering tools are not intended to produce generalized results and it is well understood that focus group results should not be generalized. However, the main purpose of this workshop was to produce a list of topics for a core curriculum that could be applicable to the larger population of indoor air scientists. A modified focus group approach worked well as a tool to facilitate this task. It provided a slightly more formal structure than the workshop's whole group interaction and gave each participant an opportunity to have input. The sessions were conducted following the ground rules of the focus group technique [3]. The participants were devided in two different groups. Each participant wrote what he or she considered to be the most important topics related to the question to be addressed. Than they presented these in turn. The discussion that followed was non-judgmental and built on each other presentation. The moderators guided the proceedings and made sure that each participant's opinions were heard. For each session, after the groups provided their input and deliberated separately, they regrouped to view, discuss and combine each other's results. The fourth component identified action plans for the implementation of the core curriculum and recommendations for future developments.

Scope and organization of the report

The scope of this report is to consolidate existing knowledge on education indoor air sciences and to provide a basis for future educational work to be conducted in a harmonized way as an international effort. The report begins with an historic perspective on how indoor air quality developed. Next it describes the trends that are currently leading education in IAS. The following part provides an overview of educational programs available at institutions of higher education, in government agencies and other professional settings. Among the programs being introduced, there are a number that propose the use of research and problem-based approaches to provide education. New approaches and emerging issues in IAS are discussed. Different players are identified and educational needs in different cultural and geographical contexts are profiled. The report ends presenting a summary and the conclusions obtained from this project.

References

[1] Moschandreas, D. J., Woods J.E., Maroni M., and Boschi N., 1995. "Evolving Educational Paradigms for Health and Building Professionals" in M. Maroni (ed.), *Proceedings of Healthy Buildings '95*, Milan (Italy), pp. 39 - 46.
[2] Boschi, N., 1999. *"Education and Training in Indoor Air Sciences"*, Kluwer Academic Publishers, The Netherlands.
[3] Krueger, R.A., 1994. *"Focus Groups: A Practical Guide for Applied Research"*, Sage Publications, Newbury Park, CA.

PART II. HISTORICAL PERSPECTIVE: DEFINING THE SCIENTIFIC FIELD

INDOOR AIR SCIENCES:
A DEFINED AREA OF STUDY OR A FIELD TO BE DEFINED

JAN SUNDELL, MSC, DMSC
Technical University of Denmark

Regarding the inadequate ventilation of schools in Stockholm: "Today we are, rightously, beginning to demand that persons working with sanitary matter should have not only technical, but also hygienic education".

Elias Heyman. Professor of Hygiene at the Karolinska Institute, Stockholm. Bidrag till kännedomen om luftens beskaffenhet i skolor. Nordiskt Medicinskt Arkiv Band XII. N:r 2. 1880.

Abstract

The need for knowledge on IAQ is primarily based on the assumption that indoor air affect humans, and that agents in indoor air can cause illness and discomfort. The question to be answered is whether there has been and today is an indoor air science? As a background is given a short review of important historical issues within IAQ and health.

Today there is mounting evidence of the importance of indoor air pollution from a public health perspective. From the history of IAQ/health can be learned that the development started in philosophy (and practice), but has during the last decades been divided into numerous specific scientific disciplines.

There is today no specific indoor air science. Scientists within IAQ/health come from a great variety of academic fields. Needed interdisciplinary projects are few. A new paradigm where generalized knowledge is as important as specific within-science knowledge is needed.

Introduction

The need for knowledge on IAQ is primarily based on the assumption that indoor air affect humans, and that agents in indoor air can cause illness and discomfort. Secondarily, indoor air is important in e.g. industrial production (e.g. cleanliness of air in pharmaceutical, and electronical industries, indoor air humidity in paper mills etc). However, in this article the focus is on non-industrial indoor air and thus "Indoor Air Science" is the science dealing with health and comfort effects of such air.

9

N. Boschi (ed.), Education and Training in Indoor Air Sciences, 9–18.
© *1999 Kluwer Academic Publishers. Printed in the Netherlands.*

The question to be answered is whether there has been and today is an indoor air science? As a background is given a short, and incomplete, review of important issues within IAQ and health. Then the question is discussed.

On the history of IAQ and health

In his large cultural geographic textbook Carter (1968) states that man's origin is in the tropical or near-tropical parts of the world. He concludes that "Man's present spread into cold climates was accomplished in spite of his physiology and was possible only because of his invention of such things as clothing, housing and the use of fire" (ibid). However, in houses and shelters not only the thermal climate is changed. The climate shell also stops the free air movement. The dilution of pollutants from close-to-man pollutant sources are diminished. Air within a shelter is always more polluted from sources such as humans and open fires than is outdoor air. This is the basis for the need of ventilation.

Ventilation comes from Latin "*ventilare*" meaning "to expose to the wind". The main purpose of buildings is to create a climate more suitable for persons and processes than the outdoor climate. Consequently, the main aim of building ventilation is to create an indoor air quality more suitable for persons and processes than what naturally occurs in the unventilated building, and to reintroduce the positive effect of being "exposed to the wind", i.a. to dilute and remove the pollutants that man himself and his activities produce.

Without ventilation, i.e. exchange of polluted indoor air with "fresh" outdoor air, any shelter quickly degrades due to the emission of pollutants from man and his activities. Provisions for ventilation have always been arranged consciously or unconsciously. Already, the Romans had ingenious solutions such as hypocausts, a kind of combined heating and ventilating system.

Throughout history man has known that polluted air may be detrimental to health. Greeks and Romans were aware of the adverse effects of polluted air in e.g. crowded cities and mines (Hippocrates, *460-377 BC*), in spite of lack of knowledge on the functioning of the lungs or the breathing.

Already in the Bible it is acknowledged that living in buildings with dampness problems is dangerous to your health. The remedies needed were quite thourough (i.a. to get rid of all affected parts of the building).

Through the medieval era little new knowledge in this field appeared. The epidemiological findings of associations between health effects and working in certain heavily polluted premises (Ramazzini, *1633-1714*), living in crowded cities, such as London (Arbuthnot, *1667-1735*, 1733), and the sad history of many young chimney sweeps (Pott, 1778) shed new light on the importance of air pollution. Later on, the death of persons imprisoned in small room volumes (Baer, 1882), or the economic burden of the deaths of slaves from suffocation during transport over the seas gave evidence on the importance of ventilation in premises mainly polluted by man. That "bad" ventilation was not only a problem in more extreme situations was also acknowledged at that time. Gauger (1714, reviewed in Bedford 1948) remarked that it was not the warmth of a room but its inequality of temperature and want of ventilation that caused numerous maladies. Bad air was held responsible for the spread of disease and for the unpleasant sensations that are experienced in badly ventilated rooms.

The general idea up to around 1800 was that breathing primarily was a way of cooling the heart - the substance of air was not required, only its coolness. But, it was also common knowledge that expired air was unfit for breathing until it had been refreshed (Wargentin, *1717-1783*, 1757). The mystery of breathing was not solved until Priestley (*1733-1804*, 1790) discovered oxygen, and von Scheele (*1742-1786*, 1777) and Lavoisier (*1743-1794*) found that air consisted of at least two gases. The role of oxygen in breathing was pointed out by Lavoisier (1781), even though Boyle (*1627-1691*)(1662), and Hooke (*1635-1703*) 100 years earlier (1667) had found that the supply of air to the lungs was essential for life, and Mayow (*1643-1678*) had discovered that there were an exchange within the lungs between the air that was breathed and the body.

Especially the work of Lavoisier (1781) was important for understanding the human metabolism, including the quantitative association between oxygen consumption and carbon dioxide (CO_2) release. During the following half century it was accepted that the concentration of CO_2 determined whether the air was fresh or stale.

Against the background of tuberculosis and other diseases known to be contracted in crowded places, John Griscom, a New York surgeon, vividly described the need for fresh air and pointed out bedrooms and dormitories as worst: "deficient ventilation ... (is) more fatal than all other causes put together" (Griscom, 1850).

Crowded rooms tend to be overheated and during this era it was considered that the discomfort in such rooms was due to excessive heat or, in accordance with the wiev of Lavoisier, due to elevated concentrations of CO_2. Pettenkofer (*1818-1901*) started lecturing on hygienic topics in Munich 1847, and installed as the first professor in hygiene worldwide 1853 noted that the unpleasant sensations of stale air were not due merely to warmth or humidity or to CO_2 or oxygen deficiency, but rather to the presence of trace quantities of organic material exhaled from the skin and the lungs (Pettenkofer 1858). He stated that "bad" indoor air *per se* did not make people sick, but that such air weakened the human resistance towards "jede art krankmachenden Agentien", i,e, in modern words acted as an adjuvant factor. In Pettenkofer's view CO_2 was not important in itself, but could be used as an indicator of the amount of other noxious substances produced by man. Pettenkofer stated "jede luft als schlect und fu*r einen beständigen Aufenthalt als untauglich zu erklären sei, welche in folge der Respiration und Perspiration der Bewohner mehr als 1.0 p.m. (*1000 ppm*) Kohlensäure enthält und".." dass eine gute Zimmerluft, in welcher der Mensch erfahrungsgemäss längere Zeit sich behaglich und wohl befinden kann, keinen höheren Kohlensäuregehalt als 0.7 p.m. (*700 ppm*) hat" (Pettenkofer, 1858). He and a number of other authors of this time suggested 1000 ppm of CO_2 (including CO_2 from ambient air) as a limit value for an adequately ventilated room (700 ppm in bedrooms), including some margin for the use of oil burners for lighting.

A number of studies of ventilation in schools, theatres, homes, etc. were conducted with the concentration of CO_2 as a measure of ventilation rate. The first Swedish professor in hygiene, Elias Heyman (*1829-1889*) at the Karolinska Institute, made an extensive study of schools with different ventilation systems in Stockholm, including measurements of CO_2, air flow rates, air temperature in the room and outdoors, and notations on number of occupants in the room as well as of speed and direction of the wind outdoors (Heyman, 1880). In schools without any ventilation he measured concentrations of CO_2 up to and over 5000 ppm, while he in schools with any kind of ventilation typically measured maximum concentrations of between 1500 and 3000 ppm. He concluded that not even one schoolroom was adequately ventilated. He also made an interesting comment on common complaints regarding "dry air" in a building with supply of heated air (up to 60° C), a sensation that he meant had nothing to do with the air humidity but rather was an effect of air pollutants drawn into the system from a neigbourhood chimney. Parallell to the complaints of "dry air" there were complaints of mucousal and skin problems, i.e. a description close to todays sick building syndrome (SBS). Also "häufige Auftreten von Kopfschmerzen" among schoolchildren were believed to be caused by "verdorbenen Luft der Klassenzimmer" (Becker, 1867, reviewed in Erismann, 1882). Heyman (1881) also studied homes and concluded that we can not rely on "natural" ventilation if we want to live in "clean" air. Wallis (*1845-1922*) in the same manner studied theatres and restaurants (Wallis, 1879). Pettenkofer and other researchers of this era, within the field of building hygiene, often stated that source control is a prerequisite for good hygiene "Ohne durch-greifende Reinlichkeit in einem Hause helfen uns alle Ventilationsvorrichtungen wenig".

Beginning with the results of Pettenkofer a number of studies were conducted between 1880-1930 in search for evidence of the toxic effects of organic substances in expired air, the anthropotoxin theory (e.g. Brown-Séguard and d'Arsonval 1887, reviewed by Bedford 1948). Since no proof of toxic effects could be found and since high concentrations of CO_2 as single pollutant caused no discomfort, the warmth of a crowded room together with smelling, but not toxic, body emissions were thought of as main sources of discomfort in rooms with bad ventilation (Flugge, 1905; Hill, 1914). Flugge (1905) wrote that the objection to an evil-smelling athmosphere was to be supported not on account of its poisonous properties, which had never been proven to exist, but on account of the resulting feeling of nausea.

Thus, ventilation was primarily a question of comfort and not of health. However, Winslow and Palmer (1915) found in a study of the effects of lack of ventilation upon the appetite for food, that there were substances present in the air of an unventilated occupied room which in some way, and without producing conscious discomfort or detectable physiological symptoms, diminished the appetite. In the experiments they controlled temperature, humidity and air movement. In later work Winslow and Herrington (1936) obtained the same results with heated house-dust taken from vacuum cleaners as source of air-pollution.

In their standard-setting work Yaglou, Coggins and Riley (1936) studied body odor in relation to ventilation rates. They stated that such odors, as a rule, are not known to be harmful. They recognised, though, that "Sensitive persons are occasionally affected in a pathological way by sitting in such rooms". Thus, again ventilation was primarily a question about comfort. ".. occupied rooms should give a favorable impression on entering, taking into consideration such factors as odors, freshness, temperature, humidity, drafts and other factors affecting the senses" (ibid). They showed that simple recirculation of air did not affect the odor strength and concluded that "from the standpoint of body odor, a room can be ventilated just as well with an outdoor air supply of 8 L/s,p as with a total supply of 15 L/s,p about 1/2 of which is recirculated. Recirculation is often desirable for adequate distribution and temperature control, but one of the disadvantages is that it smells up the ducts, fans, et., and unless the system is flushed frequently with clean air, higher air quantities will be needed to obtain satisfactory results" (ibid). They also experimented with humidifying and dehumidifying of the recirculated air and found that both technics and especially dehumidifying reduced the odor strength resulting in a reduced need of outdoor air supply for odor control.

Since the 1930's there has been only little scientific effort within the field of ventilation - IAQ - health in non-industrial premises. Odor and thermal comfort were thought of as the factors relevant in setting guidelines for ventilation.

Historically, ventilation standards have been based on the assumption that man himself is the main source of indoor pollution, mainly body odors. Health issues have not been involved, instead the classical measure of air quality has been the extent to which odor is perceived as acceptable by visitors directly on entry into the premises. This measure was used by Yaglou, and has in recent years been developed by Fanger and collegues (Fanger, 1988; Fanger at al., 1988). Fanger also, stresses the comfort aspect of indoor air pollution in stating "It is normally the perception that causes people to complain", but also "the perceived air quality ... may in many cases also provide a first indication of a possible health risk" (Fanger, 1992). Fanger has focused on the sensory load of pollution sources, besides persons, such as building materials, fixtures, fittings, furnishings and furniture.

After Yaglou'studies, the view on required ventilation levels were largely unchanged for a long time. The typical minimum value for general office spaces was 7.5 L/s,p of ventilation air, as recommended by the American Society of Heating, Refrigerating and Airconditioning Engineers (ASHRAE, 1977). As a result of the energy conservation debate in the 1970's requirements in non-smoking areas were reduced to minimum values of 2.5 (ASHRAE 1981) and 4 L/s,p (Nordic Committee on Building Regulations, NKB, 1981; Sundell, 1982), respectively. Lately, in response to increased interest in building hygiene, revised ventilation requirements have appeared with minimum values, for office spaces, raised to 10 L/s,p (ASHRAE, 1989) and about 11 L/s,p (NKB, 1991).

The problems and debate today

Environmental issues were up to 1963 primarily focused on IEQ.
With "Silent Spring" in 1963 Rachel Carson (followed by numbers of writers) changed the view on "environment" from IEQ to ambient environments "nature". Around this time reports on health problems due to the air in industries etc were more often presented, also moving the focus from non-industrial indoor air.

"Environment" were suddenly synonumous with ambient air and industrial surroundings. IAQ in non-industrial indoor environments was not on the list on environmental problems!

Not until the problems that arose with regard to radon in the late 1960's (Swedjemark, 1979), formaldehyde in the early 1970's (Andersen, 1979), house dust mites in the late 1970's (Korsgaard, 1979) and SBS in the late 1970's (WHO, 1983) did the health issue again enter the agenda regarding ventilation requirement. Typically U.S. Environmental Protection Agency (EPA) in their report on indoor air pollution research 1976 commented that "with the exception of ..carbon monoxide exposure..health effect studies have correlated health status changes with ambient (i.e., outdoor) air pollution levels, or with specific and unusual exposures of individuals to air pollution in industrial workplaces" (EPA, 1977).

During the last decade, the increased incidence and prevalence of allergy, and especially asthma, have put focus on non-industrial indoor air pollution and ventilation. Today there is mounting evidence of the importance of indoor air pollution from a public health perspective (Samet, 1993). Still, however, research on the need of ventilation during the last decades have mostly dealt with odor (Berglund and Lindvall, 1979; Cain, 1979; Fanger, 1988). Other research on ventilation has primarily been focused on thermal factors, air flow pattern within rooms (draught; efficiency of ventilation) and on energy conservation (e.g. Sandberg, 1984; Sundell, 1987).

As people spend, by far, their most time indoor in non-industrial settings IAQ and other exposures in such settings are the most important exposures related to the health of people. Lung cancer due to exposure to ETS and radon, the dramatic increase in the incidence and prevalence of allergies, at least partly attributed to IAQ, SBS and BRI mostly attributed to IAQ, airways infections associated to IAQ (but scarcely studied) are among the main health problems today related to IAQ. The health effects are seemingly primarily related to inadequate ventilation and "damp" buildings, partly due to energy conservation measures. A new trend towards green or sustainable buildings have introduced further problems due to the use of so called "natural" materials, "natural" ventilation and "old" building techniques without regarding the up to date knowledge on e.g. building physics and ventilation.

One science or many?

From the history of IAQ/health can be learned that the development started in philosophy (and practice), were divided into medicine, and technology (and practice), though in cooperation. During the last decades most scientific efforts have been in specific disciplines such as building physics, HVAC engineering, arcitecture, medicine (a number of disciplines), organic chemistry, microbiology, sociology, psychology, physics, economy etc (and practice), mostly with only small interaction. The trend has also been from an integrated approach on environment and health to a division into ambient air and health, industrial environments and health and non-industrial environments and health (and practice).

The trends has followed the general trend within sciences from philosophy (general science) into more and more specialized sciences. There is, in most academic institutions, today no place for generalists. Academic career, as well as funding of research projects, is built on further specialization within already highly specialized sciences.

There is today no indoor air science. You can't get a degree in such a discipline! In the days of Pettenkofer You could still make a career within hygiene and be specilalized in IAQ-health dealing with medical, technical, chemical and other issues. Today the basic training is within a very specialized science. And, by far most research and training related to IAQ/health is within single disciplines. Within IAQ-health related sciences typical "within-science" examples are:

- further developments in measurements of VOCs without discussing the validity of the measured VOCs in a health perspective.

- further developments regarding moisture transfer in building material, without considering the health consequences.

- further developments in calculations of air movements within a room without taking into account the disturbing factors of people (and their movements), open doors, the operation of machinery (copiers, laser printers etc), etc and without discussing the impact on human health and comfort.

Of course, within discipline science related to IAQ/health has to continue. But there is also needed much more interdisciplinary efforts. Especially within epidemiology the projects have to be truly interdisciplinary with expert knowledge within all involved disciplines. There have been only few such interdisciplinary research efforts (notably the "Danish Town Hall Study" and the Swedish "Office Illness Study of Northern Sweden").

The future

What's needed is a new paradigm where generalized knowledge (putting findings in a total perspective) is as important as specific within-science knowledge. The importance of a scientific effort should be judged in a holistic perspective.

Further is needed a more pronounced interdisciplinary approach, with top-science within all disciplines in cooperation.

References

Andersen I, Korsgaard J. 1986. Asthma and the indoor environment. Assessment of the health implications of high indoor humidity. Environment International 1986:12: 121-27.

Andersen, I., Lunqvist, G.R., Jensen, P.L., Proctor, D.F. (1974). "Human response to 78-hour exposure to dry air". *Archives of Environmental Health*, 29, 319-342.

Andersen, I. and Proctor, D.,(1982) "The fate and effects of inhaled materials". In: Proctor, D., Andersen, I., (eds), *The nose, upper airway physiology and the atmospheric environment*, Amsterdam, Elsevier biomedical.

Arbuthnot J. 1733. An essay concerning the effects of air on human bodies. London:J Tonson, 1733.

Arshad SH, Hide DW. Effect of environmental factors on the development of allergic disorders in infancy. J Allergy Clin Immunol 1992;90:235-41.

Arshad SH, Stevens M, Hide DW. 1993. The effect of genetic and environmental factors on the prevalence of allergic disorders at the age of two years. Clin Exp Allergy 1993;23:504-11.

Bedford T. 1948. Basic principles of ventilation and heating. HK Lewis &Co. Ltd, London.

Boyle R. 1662. A defence of the doctrine touching the spring and weight of the air. London: Thomas Robinson, 1662.

Brown-Séquard CE and A'Arsonval A. 1887. Demonstration de la puissance toxique des exhalations pulmonaires provenant de l'homme et du chien. Compt. rend. Soc.de Biol.,39:814.

Brunekreef B, Dockery DW, Speitzer FE, Ware JH, Spengler JD, Ferris BG. 1989. Home dampness and respiratory morbidity in children. American Review of Respiratory Disease 1989;140:1363-1367.

Cain WS. 1979. Interactions among odours, environmental factors, and ventilation. INDOOR CLIMATE, Danish Building Research Institute, Copenhagen, 1979:257-272.

Carter GF. 1968. Man and the land, A cultural geography. Holt, Rinehart and Winston. New York. 1968.

Erismann F. 1882. Die Hygiene der Schule. In Handbuch der Hygiene und der Gewerbekrankheiten. Zweiter Theil. Sociale Hygiene. 2.Abt. Verlag von FCW Vogel, Leipzig 1882:1-88.

EPA (Environmental Protection Agency). 1977. The status of indoor air pollution research 1976. Final report. EPA-600/4-77-029, May 1977.

Fanger, P. O. 1988. "Introduction of the olf and decipol units to quantify air pollution perceived by humans indoors and outdoors", Energy and Buildings, 1988;12:1-6.

Fanger PO, Lauridsen J, Bluyssen P, Clausen G. 1988. Air pollution sources in offices and assembly halls, quantified by the olf unit. Energy and Buildings, 1988;12:7-19.

Flugge C. 1905. Ueber luftverunreinigungen, wärmestauben und luftung in geschlossenen räumen. Zeitschr, f Hyg. 1905;49:363

Gauger N. 1714. La mechanique du feu. Amsterdam. H.Schelte.

Griscom JH. 1850. The uses and abuses of air: Showing its influence in sustaining life, and producing disease; with remarks on the ventilation of houses, 2nd ed. Clinton Hall, NY:J.S.Redfield.

Heyman E. 1880. Bidrag till kännedomen om luftens beskaffenhet i skolor. Nord Med Ark 1880;XII(2):1-47.(in Swedish with French summary).

Heyman E. 1881. Om luften i våra bostäder. Stockholm, Samson & Wallin. (in Swedish).

Hill L. 1914. Report on ventilation and the effect of open air and wind on the respiratory metabolism. Rep Loc Govt Bd Publ Hlth. N.S., No 100.

Lavoisier A. 1881. Second mémoire sur la transportation des animaux. Traité de Chimie, 2nd ed. Paris.

NKB, Nordic Committee on Building Regulations. 1981. Indoor climate, NKB Report No 41, May 1981, Stockholm, 76 pp.

NKB, Nordic Committee on Building Regulations, (1991) Indoor climate - air quality, NKB Publication No 61E, Nordic Committee on Building Regulations, Helsinki.

Pettenkofer M. 1858. Uber den Luftwechsel in wohngebäuden. JG Cotta'schen Buchhandlung, Munich, 1858.

18

Priestley J. 1790. Experiments and Observations on Different Kinds of Air, and other Branches of Natural Philosophy Connected with the Subject. Birmingham :Pearson New York: Kraus Reprint, 1970. (reprint from Ann Chim, 1970;7:133).

Proctor DF. 1982. "The mucocilliary sytem". In: Andersen I, Proctor DF. The nose upper airway physiology and the athmospheric environment, New York, Elsevier Biomedical, pp. 245-278.

Quackenboss JJ, Lebowitz MD, Michaud JP, Bronniman D. 1989 "Formaldehyde exposure and acute health effects study", Environment International, 15, 169-179.

Scheele CW. 1777. Chemische Abhandlung von der Luft und dem Feuer. Uppsala & Leipzig:Magnus Swederus.

Swedjemark GA. 1979. Ventilation requirements in relation to the emanation of radon from building materials. INDOOR CLIMATE, Danish Building Research Institute, Copenhagen 1979.405-421.

Wargentin P. 1757. Väder-växlings inrättningar på skepp. Kongl. Vet. Acad. Handl. 1757;XVIII:1-14. (in swedish).

Wallis C. 1879. Hygieniska undersökningar angående luftens beskaffenhet i Stockholms theatrar och kaféer. Stockholm, Centraltryckeriet. Thesis Med Faculty of Lund. (in Swedish).

WHO, World Health Organization (1983). Indoor air pollutants: exposure and health effects. EURO Reports and Studies 78. World Health Organization, Regional Office for Europe.

WHO, World Health Organization (1986) Indoor air quality research, Report on a WHO meeting. EURO reports and studies 103. Copenhagen: World Health Organization, Regional Office for Europe.

World Health Organization, (1989) "Formaldehyde", Environmental Health Criteria 89, World Health Organization, Copenhagen.

Winslow C-EA, Palmer GT. 1915. The effect upon appetite of the chemical constituents of the air of occupied rooms. Proc. Soc. Exp. Biol. & Med. 1915;12:141.

Winslow C-EA, Herrington LP. 1936. THe influence of odour on appetite. Amer.J.Hyg.1936;23:143-156.

Yaglou, C.P., Riley, E.C., Coggins, D.I., Ventilation requirements. ASHVE Transactions, vol 42, 1936, pp. 133-162.

PART III. CURRENT TRENDS IN INDOOR AIR SCIENCES EDUCATION

TOWARD AN INTERNATIONALLY HARMONIZED, MULTIPROFESSIONAL EDUCATIONAL PROGRAM IN INDOOR AIR SCIENCES: NEEDS AND STRATEGIES

M. MARONI

President of the International Society of Indoor Air Quality and Climate (ISIAQ)
University of Milano and International Centre for Pesticide Safety, Via Magenta 25, 20020 Busto Garolfo (Milan) Italy

1. Introduction

The field of indoor air sciences has great relevance for the society as the majority of the population spend most of its time indoors and the quality of the indoor environment greatly influence health, comfort, productivity and, in a word, the overall quality of life [1]. The achievement of a high quality of the indoor environment, in particular as far as air quality is concerned, depends on a number of factors: the quality of the outdoor environment; the way the buildings are designed, constructed, operated, and maintained; the activity and behavior of the occupants. All these factors in turn are linked to economic and cultural aspects specific for each community, but they are also much related to the awareness of the population and the level of education and specific competence of all the actors who play a role in the indoor environment (administrations, owners, building operators and managers, designers and constructors, hygienists, etc.). The consequence of the above is that if we want to improve and up-grade the quality of indoor air, we have to devote efforts to educate all these actors.

The educational needs vary for the various groups of people involved in the process. To simplify them into a scheme, we can consider the following main groups:
- the occupants of the buildings (i.e. the general population)
- the professionals acting for or in, buildings
- the educators inside and outside the formal school system.

This paper will concentrate in the following mainly on the education of professionals.

1. Educational Needs and Strategies for the Professionals

The field of indoor air sciences includes a wide range of disciplines and, consequently, of professions [2]. These professions have to work in an **integrated way** to achieve high quality results, that is they have to form highly integrated *"indoor air science teams"*.

N. Boschi (ed.), Education and Training in Indoor Air Sciences, 21–26.
© 1999 *Kluwer Academic Publishers. Printed in the Netherlands.*

One of the major problems in order to reach the goal of a high professional performance resides in the different approaches these professions take in their professional practice, due primarily to the different nature of their expertise and their different education.

Thus the **primary goal** of an educational strategy on indoor air sciences' is to create the basic conditions for each profession to:

- Understand the essential elements of the other disciplines involved
- Identify and maintain their own specialized expertise
- Learn how to integrate their own specialized contribution into the team in order to reach the common goal of the team's professional mission.

According to the above concepts, an **educational process** functional to this strategy has therefore:

1. to identify
- Which knowledge is "basic" and common to all the disciplines
- Which knowledge is specific of each discipline (and profession)

2. to set up an educational program for each profession aimed at the acquisition of:
- The basic tools for the "cross-professional" common understanding (core program)
- The specialized knowledge specific of the respective discipline (specific program)

1. Definition of the Core Program

The final purpose of every professional action in the field of indoor air is to create a high quality of the air people breath in the confined space of concern. The quality of indoor air in a given space is actually the net result of:

- the actions leading to air deterioration (pollution)
- the actions leading to air amelioration

which together can be defined as the **indoor air process.**

It can be considered to be "basic" for every discipline to understand:

- the fundamental rules of this process (model of the process)
- the actors that play a role in the process (sources, emissions, pollutants, ventilation, etc.)
- the consequences of the process (on ecology, economy, health, social behavior, etc.)
- the tools available to modify, or intervene on, the process (planning, design, source control, air monitoring, diagnostics, HVAC, maintenance, cleaning, management, information, etc.)
- the disciplines (and the related professionals) that use or can use these tools, and their possibilities and limitations
- the implications for the society of indoor air, both in terms of indoor air management or mis-management (indoor air policies, consumer goods production policies, energy policies, ecological policies, social policies, etc.).

Among the others, one of the essential goal of such a core program is to provide to the future professionals a **common dictionary of terminology** that should enable them to understand the meaning of the essential words used to describe, analyze and intervene on, the indoor air process.

This apparently obvious achievement is far from simple to obtain, since each of the words (and the underlying concepts they express) is in its turn linked to other basic knowledge pertaining to the original field from which it derives, that the trainee may not necessarily know and appreciate in its complexity. However a balance can be achieved and it is the responsibility of the educators to appreciate the starting level of the trainees, their possibility to understand the messages properly, and the possible needs of background details functional to their comprehension.

1. Definition of the Specialized Program

By definition, the specialized program is specific to each profession and therefore has to be specifically devised for each professional figure. The variety and type of professional figures active in the indoor air field vary in the various countries and economic and cultural contexts. This would implicate that a closed list of professional labels valid everywhere does not exist, although one may recognize some exceptions. A further complication is that similar labels may have a quite different professional practice (and level and contents of education) in different places.

If the above assumptions are true, then their natural corollary is that the definition of the professions – and of their professional practice – has to be made case by case, or country by country, or context by context.

A rationale process toward the definition of the professional profile of each specialist should include a thorough analysis of:

- what each professional figure in a given context **actually does**
- what each professional figure in a given context **ought to do**.

Obviously, these two elements not necessarily coincide. Indeed, the existence of an "ought to do" different from an "actually does", is the justification and the reason of existence of an educational strategy innovative with respect to that currently in place in our educational Institutions.

Once the ought-to-do is defined, the objective of the educational process can be analyzed and programmed in terms of three basic operational perspectives of the professional profile:

- to know
- to be able to do
- to be able to be

To Know. The first dimension of the profile pertains to the indispensable knowledge the professional must have in its specific field of activity. This knowledge can be indispensable either to allow him/her to understand the problems to be solved and the context in which they are located or to allow him/her to recognize the potential (and the limitations) his/her discipline has to handle the proposed situations.

In general, the educational programs of any profession consider this objective the "theoretical" component of the education and address this need with lectures, personal study, conference attendance, etc.

To Be Able To Do. This dimension of the professional profile is very important and often difficult to implement. In the various cultures, the need for a professional to be able to do practical activities in first person (as opposed to simply know in theory how to do them) is differently appreciated. Many societies accuse the formal school process to generate professionals with scarce and insufficient practical skill, though they may have an excellent theoretical formation. These accusations to the school systems would advocate for more "practical exercise" or "practical problem solving" capacity; however these accusations tend to neglect the fact that a practical, inductive process of learning (from the single case to the general theory) does not produce a systematic learning and may result severely inadequate in front of a very complex reality that is under continuous change. Under such circumstances the trainee is often left disarmed and sometimes unable to adapt to new solutions.

An apparent good solution should be to achieve an equilibrated balance between theory and practice. This would imply a clear identification of the *essential prototypical actions* the trainee will encounter in his/her future work and the inclusion of practical applications of them in the educational program. A tutorial discussion on these exercises can help the trainee generalize their value and understand how diversified the reality will present these tasks.

To Be Able To Be. This professional dimension is very frequently ignored or undervalued in most educational programs. In fact the indoor air professional has to work in a team and often has to relate to a number of other interlocutors. The way these relations are handled is very often essential to the work outcome at least as the specific technical skill necessary to perform the work. The difficulty in this part of the educational program resides in the diverse disciplines necessary to be activated in order to construct the ability of the trainee. Some of these disciplines may be very remote from the specific field of the professional and often perceived to be "waist of time". Moreover the definition of the specific educational needs requires a full and correct analysis of the work context and the social context, which determine rules of behavior, roles, social values, and other important aspects.

1. The Delivery of Education in Indoor Air Sciences

The principal institution involved in the delivery of education is represented by the formal school system. Although a basic level of education on indoor air sciences should also be part of the primary and secondary levels of education, the University is the level at which professionals are specifically educated and formed into their specific jobs and missions.

The four main streams of professional education offered at University level related to buildings consist of architecture, engineering, biological sciences, and health sciences (medicine, public health, nursing). The academic curricula of these professional streams are rather different across the world. Moreover, in the various countries several degrees are often offered for these professions, ranging from graduate to doctoral level.

In the lack of an international standardization, or at least harmonization, in the basic curricula, it is difficult to envisage a single curriculum or core program on indoor air sciences that would easily fit as a package into every existing academic program. In practice, then, the insertion of a number of "new " subjects in the curriculum would necessarily require the elimination of other topics, given that the academic programs in the Universities are usually already taking a full time commitment of the students. In spite of these difficulties, we believe it is necessary to try to develop a "core program" and to propose to the Academic managing institutions to consider the insertion in the early programs of these four educational lines of those specific subjects pertaining to other disciplines that are not covered by the existing teaching programs. In most cases, this can be achieved by calling in faculties from other academic schools to complement the basic teaching provided by their own faculties. Such a multidisciplinary approach will inevitably offer new stimulating experiences also to the body of inner faculties who will be exposed to new thinking and ideas.

2. Conclusions

There is a clear need for a multidisciplinary education in indoor air sciences that should be harmonized at international level. Due to differences across countries and cultures, such a harmonization can be achieved primarily at the basic university education of the four main professional programs targeted to architects, engineers, biological science graduates, and health professionals. The process of harmonization will necessarily proceed gradually and will encounter resistance and difficulties. However, the great benefit expected at society level by this educational strategy should be an argument strong enough to convince politicians, education planners, professional associations, and the general public about its convenience and urgency.

26

3. References

1. Maroni, M. and Berry, M.A. (eds.) (1989) *The Implications of Indoor Air Quality for Modern Society*. Pilot Study on Indoor Air Quality. NATO/CCMS Report on a Meeting in Erice, 13-17 February 1989.

2. Maroni, M., Seifert, B., Lindvall, T. (eds.) (1995) *Indoor air quality – A Comprehensive Reference Book*. Air Quality Monographs, Elsevier Science, The Netherlands.

TRENDS IN THE POST GRADUATE EDUCATIONAL CURRICULUM OF THE INDOOR AIR SCIENCES (IAS)

LARS MØLHAVE, PH.D.
Department of Occupational & Environmental medicine
University of Aarhus, Denmark

Abstract

This paper discusses the activities performed by the professionals in indoor air sciences and takes it starting point in the 30 years long case of formaldehyde in the indoor air as a typical example. The case illustrates the many types of functioning and operating within the multi-disciplinary society of indoor air sciences (IAS).

In IAS there is a need not only for interdisciplinary cooperation but also for each individual scientist to have multiple training. Only in this way will the professionals in IAS merge into an IAS society.

The who's, where's, when's, and how's of such a training is discussed. On this basis some of the habits and prejudgments preventing interdisciplinary communication and progress are identified and discussed.

Major reasons for the lack of cross-disciplinary communication and understanding are the lack of an agreement on common goals, ethics and responsibilities and of a common vocabulary. The author's personal definition of human rights and responsibilities in relation to indoor air are presented and the establishment of a task force on an IAS vocabulary is suggested. Finally, the consequences for the curriculum of a potential post graduate education on IAS is discussed.

Keywords

Education, training, IA sciences, research, post graduate education, health, exposures, air pollution, building evaluation, indoor environment, ethics.

27

N. Boschi (ed.), Education and Training in Indoor Air Sciences, 27–43.
© 1999 *Kluwer Academic Publishers. Printed in the Netherlands.*

Introduction

The group of professionals working within indoor air sciences (IAS) appears to be fragmented into experts fields with only partly overlapping interests. Few experts in IAS do whole-heartedly say "I am a scientist or a professional in IAS with main functions within the IAS and have a background in many types of educations or professions. Instead many of the IAS professionals have a firm basis in their original education and only half-heartedly accept that they have to bend to other educations with different traditions and technical language.

This paper advocates that the overall theme of an IAS postgraduate educational curriculum should be to give existing and future IAS professionals (i.e. IAS post graduate students) the feeling that they belong to a cooperating IAS society.

The aim is to inspire to that courses and classes have a common aim and super structure, and are not just focused on specific indoor air problems viewed unilaterally e.g. from an engineering or a health perspective. Such a curriculum might merely duplicate already existing activities in these professional societies.

Therefore, it is discussed how the IAS professionals can make a difference by doing cross disciplinary research in a professional discipline called e.g. the sciences of indoor air (IA). This discussion hopefully will show the contents of an IAS postgraduate educational curriculum. Special focus is on the trends in the changes of the curriculum of the past caused by changes in the surrounding society or in our life styles.

The results and tools in IAS are of interest for many groups dealing with indoor air. This presentation mainly discusses the education of experts or professionals fully or partly working within IAS.

It must be kept in mind that this presentation is meant as an inspiration to further discussion and is a biased simplification based on the author's personal experiences. Because of this all references have been left out of the paper. Also it is mainly prepared as seen from an European point of view.

Fragmentation of the IA Research

THE MIXED GROUP OF IAS PROFESSIONALS

Table 1 shows the history of the formaldehyde case. Table 2 shows in this case the functions performed by the members of the IAS society, their original types of educations and the users of their services. They fall in four groups of experts: source control (constructors or producers), exposure control (HVAC engineers, hygienists), health effects (health professionals, toxicologists) and public health (regulators, guideline setters).

The group of IAS professionals ranges from non-IA specialists (producers, IA consultants, building constructors, or health professionals) with educations (e.g. technical or medical) which have other main functions than IA and who do IA research within their own special discipline using results and tools from IAS, to dedicated IA researchers with a multiple educational background and who are producing the tools and results used by others and who are mediating research to the other researchers.

The professionals using the results and tools of IAS: In table 2 some expert educations show up in both columns which e.g. indicate that there are engineers with no special interest in IAS but who are performing some of their work as routine e.g. in operation or maintenance of buildings.

Most of these professionals feel more attached to their original expert education and their own traditional professional societies and IAS is only one of many factors for them to deal with. They do research within their own societies and rely on the results of IAS. They have their own societies to organize special courses in IAS seen from their perspective. It can be questioned if they are "true" IAS scientists but not that IAS scientists have a role to play in their education.

The IAS scientists: A group of cooperating IAS scientists of mixed expert educations exists in table 2. In addition to their original education they have professional experiences which allow them to bridge health, technology, exposure measurements, building sciences etc. They, therefore, have a multi disciplinary background and are involved in several types of functioning including mediation of research between other disciplines.

This mixed group of IA scientists has no identity or name of their own. In this presentation they are called the IAS professionals or IAS scientists.

The non-expert user group: In table 2 a non-expert group of users is identified. They have many different types of education ranging from preschool to expert education. Some of them may have a very personal (e.g. health) reason for their association with IAS matters. Obviously there is a need for education here which IAS scientists may participate in. However, that education is not the aim of this paper.

THE COMPLICATIONS OF WORKING AS AN IAS SCIENTISTS

divergences within the group: The IAS scientists are often said to make life more complicated for the non-IAS scientists e.g. by claiming that engineers are causing many of the indoor air problems because of their too restricted view on indoor air relations and ventilation or that medical doctors overlook the health problems by focusing on adverse body defects and body related causes of diseases thus neglecting environmental exposures. Consequently, some IAS scientists are considered to be troublemakers by others because they force them to consider more variables outside their expertise. We have to learn to communicate positively, to accept the limitations of other professionals with other approaches and tools and to understand that a positive contribution in this sense is one of the justifications for talking about special training of IAS scientists.

The scientific standard: The IAS professionals' discussions especially between engineers or health professionals show that IAS professionals do not talk the same technical language or have the same vocabulary. In addition we do not have the same overall goals for our functions as IAS professionals and have not had the time or will to define an overall strategy and aim as a society of IAS professionals.

Researchers from other research disciplines point at the fact that many IAS scientists are not truly unbiased by e.g. economical or personal influences thus indicating the need of an ethical code of IAS scientists.

We deal with cases and emergency situations. Mainly, the IAS scientists or hygienists solve acute problems for surrounding society but IAS sciences do not show optimal efficiency. This is illustrated by the formaldehyde case which took the IAS researchers too long time to solve.

IAS research lack focused research. Formaldehyde could have been used as a model pollutant and thus we could have learnt more which by generalization might have prevented similar problems from appearing later. Therefore, we need better research techniques and information procedures.

We need better publication practice. Few publications have sufficient scientific quality. Many of the published papers represent ad hoc craftsmanship in dealing with indoor air problems but add little to our basic knowledge of general processes and factors in the field of IAS.

The incomplete group: In addition to the background knowledge mentioned above, the optimal IAQ solutions involve factors related to psychology, economy, the present technical possibilities etc. One consequence of this is that there is no absolute optimal solution to the functions of an IAS professional. Another consequence is that in the future we probably will have to find optimal IAQ solutions with additional considerations of economical, political, psychological, sociological, etc. factors. It follows that the IAS-group lack members with both a sufficient understanding of indoor air problems and terminology and with cross-disciplinary experiences from some of these essential contributing educations (e.g. psychologists, toxicologists or economists). Therefore, more types of educational background must be added to the IA group of scientists and the basic IAS training.

Existing and future IAS scientists belonging to all these groups are the targets for a planned IAS education. Therefore, there should not be a lack of students but when we consider the many problems described above do we then really want to establish an IAS advanced IAS education or should we continue to rely on ad hoc solutions as hitherto?

If we accept to establish a specialized IAS education we have two options:

1. we can consider the IAS professionals as belonging to each own's original educational society and form ad hoc IAS-groups for each case/study. The IAS training may then best be done within their original societies.

2. The problems IAS professionals deal with are generally multi-disciplinary. Therefore, generally only cross-disciplinary solutions or research are useful. Further multi-trained persons are more trained in establishing the required communication to and between bordering disciplines. Further more, we need scientists with training in the identification of future IAS problems to move IAS out of the present case-type of working. In conclusion, we may accept the need of specialists with this special multi-disciplinary training and mediator function. These persons are the potential members of the IAS society of IAS professionals and are the targets for the advanced education discussed in this paper.

The functions of the IAS professional: As illustrated with the formaldehyde case the purpose of the IAS scientists is to be multi-disciplinary and to do research, education, IA-consultancy etc. In their work with an IAQ problem the IAS researcher will be dealing with complex systems/problems not only to solve an actual and real problem but also to improve the general knowledge of the processes involved.

The research functions of the IAS scientists include as illustrated in the formaldehyde case:

- Development of the research tools needed by other IAS and non-IAS professionals.
- Identification of potential future IAQ problems.
- Documentation of causalities, exposure-effect relations, related effects (such as economical), etc.
- Assistance in regulation and standardization.
- Assistance in product development and changes of codes.
- Development of models for or descriptions of processes dealt with in IAS sciences.
- Creation of tools, procedures, and methods which other can use for improvements of IAQ in mitigation, construction, maintenance, etc.
- Creation of the back ground procedures and data needed for guidelines etc.

The advanced educational functions in IAS include these topics. In addition the education as indicated previously should include:

- Promotion of ideas of the acceptable IAQ,
- The use of a common set of available tools , etc.
- Pointing on who are responsible for good IAQ in a building.

When this IA research has been done (as in the formaldehyde case), other non-IAS professions take over the subsequent routine work (e.g. constructors, HVAC engineers or physicians). Formaldehyde still occupies these other professional groups but not the IAS professionals or researchers. On the other hand we have to give these other groups guidance and teach them about the multifactorial complexity of the problem. As explained above they need another and otherwise specialized education.

Two educational functions therefore exist: training and education of IAS scientists, and training and communication to non-IAS scientists, consumers, occupants, MSs, engineers, etc. Only the first of these is the target for this discussion.

The IAS Society and Its Functions

HOW IS A PROFESSIONAL IA SOCIETY DEFINED?

As we see it today IAS deal with health, economical, or environmental consequences of exposure factors related to indoor air and strive to improve human health and environmental conditions in non-industrial buildings by optimizing exposure factors related to indoor air. IAS try to improve human health and to reduce economical or environmental consequences of exposure factors related to indoor air in non-industrial buildings.

The main functions of the IAS professionals are as explained above: research, education, mitigation, operation, and maintenance. The IA sciences generally deal with complex IA problems with no known origin or causality. They are typically confronted with low level exposures, (meaning exposures at a level where causal effects are normally insignificant and measuring methods are lacking) or with multiple exposures' scenario in which interactions can not be predicted or handled.

When the IAS scientists have passed the steps illustrated by the formaldehyde case and made handling of similar situations routine, other professionals take over this subsequent routine work (e.g. engineers or physicians). On the other hand we have to give these other groups guidance and to teach them about the multi-factorial complexity of the IA problems and how IAS professionals can help them.

Among the things which could make the IAS professionals feel as members of an IAS society are that they have the following in common:

> Definition of the working and research fields.
> Tools for research or investigations.
> Channels of communication/language.
> Common code of behavior, ethics, or goals in:
>> - Research and internally among colleagues.
>> - Public/consultancy/professional IAS work.
> Plan for recruiting young persons.
> Education plans pregraduate and post.

A common history and "Old saints"

Presently, there is a lack of agreement on most of these issues, which therefore will be discussed in the following.

The IAS Research Universe

VARIABLES IN INDOOR AIR SCIENCES

The research universe of IAS consists of a selection of all the processes and variables which are used for the description of the entire universe. In principle this selection or grouping identify those variables or processes which experiences teach us we must consider in descriptions of the indoor environment and those which we know are irrelevant of IAS. However, a group of intermediate factors or processes exists which cannot be classified as entirely irrelevant or relevant. For some of them the reason is lack of knowledge and for other it may depend on the situation if their influence is or is not inferior compared to other factors.

It is a main purpose of IAS to develop the knowledge, which allows a better understanding of when and where these intermediate variables or processes are relevant. This is often done in an iterative process which leads to refinements of the description of the research universe and to developments of models which allow better predictions.

Figure 1· shows the traditional and simplified IAS grouping of the IAS relevant variables into four subgroups of related variables: source, exposure, and effects variables, plus an additional group of variables called response modifying variables. Arrows between these groups of variables indicate processes or interactions, which associate the variables. Overlapping areas indicate variables, which (as odor) may be e.g. an effect or an exposure measure. The figure reflects the existing expectations for a general IAS universe and includes only those factors and interactions which significantly affect the relevant outcome variables and reject constant factors or factors with negligible effects on the research universe. This reduces the number of factors and interactions to deal with and describes the interactions or mechanisms with a minimum of complexity. However, the gained simplicity is contra weighted by lost accuracy and the purpose of the IAS science is to develop a more accurate grouping of the variables and the description of the mechanisms without loosing too much simplicity.

FIGURE 1: The simplified IAS research universe

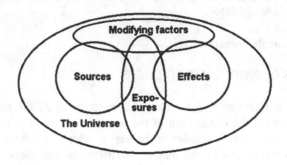

The traditional IA research universe

<u>The group of sources of exposures</u>: includes source related variables such as source strength, source type, amount, age. The interactions or processes related to this group could be mutual reactions between sources which are in contact, reactions between sources, and the buildings' atmosphere, chemical reactions in the source, evaporation, microbial growth in the source, etc.

<u>Building, atmosphere, and environment</u>: Traditionally IAS include such building variables as room size, ventilation, air movements, air concentrations, climatic factors, - technical installations, insolation, concentrations of air pollutants, etc. The main factors are air borne chemicals or biocontaminants. Typical interactions or processes are adsorption, emission, dilution by ventilation, etc. and uptake or emission from persons or targets directly.

<u>The targets</u>: Traditionally this group of variables includes factors from human health science and toxicology. Examples are target tissues, type of effects, internal transport in the body, metabolism, duration of occupation, respiration, gender, health status, etc. The interactions include emission or absorption of pollutants from the atmosphere, distribution or metabolitization of pollutants, exchange of pollutants within target tissues, etc.

<u>The group of interacting variables</u>: these groups of variables are called response-modifying variables because they are not directly responsible for the causality but merely modify the outcome. The group includes variables which to some degree affect emission from sources (e.g. variations in air pressure), spreading of pollutants indoors (e.g. wind

direction outdoors) or human responses to exposures (e.g. coexisting medical conditions). It is obvious that these factors' associations to this group of interacting factors are depending on the situation and it is one of the purposes of IAS sciences to minimize the number of variables in this group without loosing accuracy or increasing the complexity beyond acceptable limits.

Research

SCIENTIFIC ACHIEVEMENTS HAVE THREE ASPECTS

From the discussion above it follows that one of the aims of IAS research is in an iterative process to refine the description (i.e. which factors to include or how to group them) of the research universe and to develop models which allow better predictions. Three aspects of research or scientific achievements have been identified. They are the positivistic aspect, the hermeneutic aspect, and emancipatory aspect.

It has been postulated that those who in positivistic research build models are craftsmen. The true research according to this point of view is the realization prior to building the model: that it is needed.

THE POSITIVISTIC ASPECT OF IA RESEARCH

This is the traditional research type of natural sciences. Most engineering, IA-consulting work, or air quality measurements fall into this group. In this research the causalities, processes, or interactions between subsystems of the research universe and internally within the subsystems are identified and described or approximated mathematically. The aim is to make predictions of the changes expected to follow from disturbances of the research universe.

The positivistic research process often happens in iterative steps such as establishment of theories/models, challenges of the theories, refinements of theories, new challenges, etc.

THE HERMENEUTIC ASPECT OF IA RESEARCH

It is a hermeneutic research process to find the possible solution of a problem or to come to the conclusion that a certain topic should be taken up for further positivistic research. Not all scientific achievements are based on quantifiable variables and the causality may be of the form: "I went on vacation because I was tired (or had the SBS)". This research aims at understanding why other persons do as they do. Investigations of this type are called hermeneutic research and may e.g. use categorical or statistical procedures to describe responses of groups or individual persons. Hermeneutic research can be exemplified by many types of ratings of subjective symptoms, stress or psychological factors in IAS research.

The tools of such hermeneutic research are the backbone of much of the IA-research although most IA-scientists come from a positivistic background and have little training in hermeneutic research. Training here is urgently needed in IAS. This is especially the case for psychological input which is very important in explanations of IA problems.

EMANCIPATORY ASPECT OF IA RESEARCH

The emancipatory aspect of research is research of the habits, norms or prejudgments e.g. preventing appropriate positivistic or hermeneutic research and preventing adoption of the results of such research in society as a whole or inside IAS. The emancipatory research is aimed at understanding why we our self do as we do and at shifts of paradigms. The tool is often reflection and discussions for identification of reasons for the building of barriers between groups, which would benefit from interacting positively.

Researchers try to make rational research. They try to select the optimal research tools based on their own best professional judgements. In a borderline research area such as IAS subcultures of researchers meet and interact but the "best professional judgments" and the definitions of rational research differs among the subcultures of IAS researchers.

Examples:

1. This NATO workshop is a piece of emancipatory research of the factors, which prevent us from doing better positivistic or hermeneutic research.

2. It is emancipatory research to investigate the factors and mechanisms which are blocking for the implementation of existing knowledge into renovation and construction of buildings, why occupants, building owners' and regulators often find it difficult to communicate constructively, or why engineers and health sciences in IAS communicate so poorly.

3. The absence of economical or social/psychological professionals in most parts of IA research has no rational explanation. Why is it so? Such lack of acknowledgments of the relevance of other researchers' activity is blocking for developments of the IAS research area and prevents it from being respected and acknowledged as a relevant and independent research area.

Traditionally, IAS have focused on learning from past mistakes. To become a scientific field it must learn to make models for the future and these models should not (as presently) be on details of existing problems but also predict where new constellations may cause new types of problems in indoor air. Only by introducing emancipatoric thinking in the IAS field of research do we develop IAS to be a scientific field of its own rights.

THE FUTURE IAS RESEARCH UNIVERSE

In addition to the traditional variables and interactions, new factors can be expected to emerge on the IAS scene. We need procedures, which allow us to identify these. The IAS scientist will have to learn that the IAS universe will be still more multifactorial and may come to include many new factors previously considered inferior to IAS.

Interactions: The reasons of such changes in our society are an increasing sensitivity/awareness of IAQ among the occupants, new techniques and materials that causes new problems, interactions between exposures which require more complex models for the explanation, or increased requirements of efficiency in use of resources.

Sources of exposures: Examples of new additional IA sources to consider are noise, thermal loads, illumination, etc. In addition social and psychological factors may become important for the evaluation of human responses to indoor air quality.

Building, atmosphere, and environment: New atmospheric factors may become relevant such as air temperature, noise levels, stress/strain, and illumination. In addition, other exposure routes will become relevant such as contact exposures (e.g. to sedimented dust) or social interactions among occupants.

The targets: In the future IA sciences should include new effect/outcome or new effect modifying factors related to IA from e.g. economical, political, or regulatory sciences, or from environmental, biological sciences or psychology and sociology.

Also other targets of exposure will be relevant such as environmental conditions, buildings, furniture, instruments.

In conclusion, the last IAS case has not yet arrived. On the contrary, the number of cases can be expected to increase and in contrast to the present situation we must train ourselves in how to identify and handle new problems with unknown origin and not focus on how to handle old types of IA-problems. Our working field yesterday is routine jobs for e.g. engineers and health professionals tomorrow.

Ethics and Moral Codes

Previously, it was pointed out that IAS need a common moral code of ethics. A generally accepted example therefore cannot be given. For this discussion I have tried to formulate fair ethical goals for my own research in the following statements (the draft Mølhave version by 1998):

Statement 1: As a principle of precaution, exposures are unacceptable if they do not belong to the natural background or have no obvious reason or accepted purpose.

Comments: We all have a responsibility of not harming or offending our neighbors if it can be avoided. However, in the same sense we must accept some exposures from them if they are doing what they can to avoid exposures of us to the level of harm.

Statement 2: From an ethical point of view everyone has the right to know if they are exposed to unacceptable exposures beyond what could be expected in the situation obvious to them.

Statement 3: The WHO definition of health (WHO 1947) applies for indoor air. It states that health is more than the absence of diseases but also includes optimal functioning and comfort.

Comments: Exposure to pollutants which significantly decrease the health, functioning, or comfort of occupants is, therefore, unacceptable.

Statement 4: Every person has the right to an acceptable IAQ (WHO: Agenda 21, equity principle).

Comments: In this context unwanted or unacceptable exposures are defined as the presence of indoor exposure factors at levels which cause unwanted or unacceptable effects on the occupants.

Statement 5: Every one has the right that other respect their integrity and their evaluation of exposures (Draft UN declaration on human responsibilities 1997).

Comments: As the evaluation of the optimal indoor climate differs from person to person the possibility of some level of individual control of indoor climate etc. is, therefore, mandatory.

Statement 6: The optimal indoor air quality can be defined using both human health and environmental factors. However, the concern of an optimal environment should have the long and short-term effects on the human health as first priority.

Statement 7: Economical factors can be used as indicators of human health, performance and comfort or as indicators in a political prioritization of a broad spectrum of social or - economical activities. When used in this last meaning they should not have priority over human health, performance, and comfort.

Statement 8: Whether private or governmental, no group, organization etc. associated with a building stands above the responsibility to advocate or work for acceptable air quality for the occupants (draft UN declaration on human responsibilities 1997)

A Vocabulary of the Indoor Air Sciences

As pointed out above IAS cannot function without a generally accepted common vocabulary. An IAS task force should be established to define a common interdisciplinary language for professionals working in IAS. In addition, such a vocabulary would facilitate text book writing and education in all IAS working fields. The aim of this vocabulary should be to communicate definitions interdisciplinary, but not to make in-detail scientific descriptions.

This work on the vocabulary would inevitably find conflicts between established phrases in associated research fields. It would be a much-needed IAS emancipatoric research to identify and settle such conflicts which affects the work and communication of IAS scientists to other disciplines.

The IAS Curriculum

THE AIM OF THE EDUCATION

The candidates: The candidates as IAS professionals may come with many types of background education. The main target for the education should be IAS professionals with an expert education and with a major part of their work related to IAS. This would typically be IA consultants, building engineers, building constructors, building operators, chemists, physicists, microbiologist, occupational or environmental hygienists, health professionals, psychologists or sociologists, toxicologists, lawyers or regulators.

The focus is not on routine work, which other professions take over subsequently (e.g. HVAC engineers or physicians). However, we must inform and teach other than IA researchers that we actually can help them.

The education must give these students an IAS relevant background knowledge, training and skills in IAS, attitudes and motivation. Examples of these are shown in table 3.
Education of IAS researchers must make them multi-disciplinary. As illustrated with the formaldehyde case the education should train the students in all the steps from source, emission, environmental concentrations, exposures/uptake, effects, to evaluations and guideline setting.

Especially each IAS relevant type of IAS functioning must be trained both old and new. The education must define the research universe by its variables and interactions etc. and must open the IAS society for new additional IA variables and procedures. In the future we will have to find optimal IAQ solutions with additional considerations of economical, political, psychological, sociological, etc. factors. The simple research universe will not be adequate in the future.

The exposure evaluations must include other exposure routes than the air borne and other targets than the human body.

Guideline setting and regulation is not the main focus of IA sciences but an IA scientist must know how to advise in regulation, standard setting, product developments, etc.

The motivation of the students should make it easier for them to feel that they belong to an IAS society and are IA scientists. It is easier for all of us to stick to our own traditional professional societies and have them to organize special courses in IAS as seen from their perspective. Therefore, the advanced education should underline the need for cooperation and the removal of the fear of entering other fields of expertise (special technical classes for MDs and visa versa?).

TABLE 1: The typical IAS case: 30 years of formaldehyde problems: the Danish version

---

1965-1970 An IAQ problem emerges.
- Complaints, certain buildings.
- Legal, regulatory, and economical conflicts.

1970-72 Suspicion on formaldehyde as cause.

1971-73 Measurements find formaldehyde in indoor air.
- U-F glue suggested as a possible source.

1973-77 Formaldehyde as source is confirmed.
- Official regulation on emission, construction, and production.
- The first legal and economical fights are settled.
- Better mitigation procedures are established.
- New U-F products are developed.

1977-78 Sensory effects of formaldehyde are investigated.
- New NOEL, LOEL, and dose- response relations are established.
- The official regulation is revised.

1980-85 Exposure guidelines set.

1985-93 Concern about adverse effects of formaldehyde in animal studies are confirmed in epidemiological studies.

1993-95 Guidelines are revised.

1995 The formaldehyde problem is history in IA science but still occupies producers, engineers etc.

TABLE 2: The educational keywords from the Formaldehyde case

Type of activity. Keywords for a curriculum.	The IAS professionals. Potential students.	The users of the services.
Registration of complaints and health effects	Health professionals: - GPs, - Occupat. Med. - Hospitals Hygienists	Occupants, Regulators.
Building: - construction - operation - maintenance - mitigation	Engineers Constructors Building operators IA-consultants	Building owners Occupants
Guidelines Regulation	Engineers Lawyers/regulators Health professionals Toxicologists	Engineers IA-consultants Building owners Producers
Air quality control Exposure assessments	Hygienists Building operators	Health Professionals Building operators Hygienists Producers Owners
Treatment of patients	Health professionals	Occupants Building owners
Toxicological evaluations and tests	Toxicologists Health professionals	Regulators Health professionals Hygienists
Production and product development	Producers Engineers IA-consultants	Engineers Constructors Consumers

TABLE 3: Keywords of an IAS curriculum for postgraduate education

Pollutants and other exposures:
- Pollutants' physics and chemistry.
- Toxicology of pollutants.
- Measurements of the traditional variables related to source, emission, environmental concentrations, etc.
Identification of future new types of harmful pollutants.

Sources and materials:
- Physics of source emissions.
- Humidity in buildings/thermal environment.
- Adsorption/absorption in materials.
- Principles of product development.
- Building techniques.
- Measurements of the traditional variables related to source, emission, environmental concentrations, etc.

Microbiology:
- Basic biology of microorganisms.
- Humidity in buildings/thermal environment.
- Biological growth as source of pollutants.

Buildings and the indoor environment:
- Ventilation/HVAC.
- Building sciences.
- House keeping practice.
- Ways of mitigation, operation, and maintenance.
- Present techniques for increasing the efficiency in the use of resources.
- Measurements of the traditional variables related to transport of pollutants within buildings.

Human physiology and psychology:
- Physiology of main human targets.
- Uptake in humans, metabolism, and detoxification.
- Endpoint effects.
- Psychology/sociology in relation to environmental exposures.
- How the sensitivity of the occupants can be measured and how it affects the occupants' responses.
- Measurements of the traditional variables related to exposures/uptake, effects, etc.

Public health topics:
- Toxicological evaluations.
- Exposure assessments.
- Dose estimations.

- Epidemiology.
- Documentation of causalities, exposure-effect relations, prevalence-related effects (such as economical), etc.
- Monitoration of social/technological trends in:
    - Changed sensitivity of the occupants.
    - New materials and building techniques.
    - Demand for increasing the efficiency in the use of resources.
- Predictions of potential:
    - Old types of IA problems
    - New problems caused by society and technological developments.
- Using toxicological data for IA evaluations.

Legislation and regulation:
- Laws and responsibilities in relation to indoor air problems.
- Principles of regulation. Standardization, guideline setting whether these are focussed on optimization of health, economics or are otherwise focused.

Other topics:
- The common tools of research, mitigation, etc. (positivistic, hermeneutic, and emancipatoric).
- Publication practice and common channels of communication. (Indoor Air, Conferences, ISIAQ, etc)
-The common IA language, terms, abbreviations etc.
- The purpose of the IA research which is not merely the control of the level of pollution and optimization of it according to guidelines set by others. This is engineering science. IA sciences are not merely treatment of occupants, health effects caused by indoor air. This is medical science.
- The common goals and ethics. Each teacher should use the same set of ethical rules and goals when preparing the classes.
- The history of IAS.

PART IV. ON-GOING EDUCATIONAL PROGRAMS

INDOOR AIR SCIENCE TRAINING AND EDUCATION IN FINLAND

P. J. KALLIOKOSKI
University of Kuopio, Finland

1. Abstract

Extensive indoor air research has been established in Finland (population 5 million), mainly during the last ten years. Simultaneously, several training and education programs has been launched in this field. The main actors in this progress have been the research teams in Helsinki and Kuopio. The former consists mainly of the researchers in Helsinki University of Technology (HUT) and State Research Centre (VTT) and it works in close collaboration with the Finnish Society of Indoor Air Quality and Climate (FISIAQ). The latter includes the researchers in University of Kuopio and its hospital as well as in the units of the national Public Health Institute (NPHI) and the Finnish Institute of Occupational Health (FIOH) in Kuopio. Both research agglomerates have a lot of cooperation. In addition, the main unit of the Institute of Occupational Health in Helsinki has provided training in indoor air quality problems at workplaces. Most recently, the university hospitals in Helsinki and Kuopio have established indoor air clinics for people who suspect to suffer from building-related illnesses. These clinics also provide basic information about these issues for their patients.

2. Historical background

When the training started at university level in middle 80ɔs, a medical state advisory committee had claimed that mechanical ventilation was the main culprit for the building-related health problems. Tight building envelope was naimed as a contributing factor. These allegations were based on the results of the first British sick building study. Therefore, it was natural that the role of ventilation for the symptoms and perceived air quality became the first target of indoor air research in Finland. The research was initiated as collaboration between the HVAC laboratory of HUT and the University of Kuopio. This cooperation has lasted ever since and many other research institutes have joined to it later (the most important ones have been mentioned already above). The role of the HUT was natural because it was (and still is) largely responsible for training and research in ventilation engineering. The reason for the participation of the University of Kuopio was its experience in studying mold problems in agriculture (in collaboration with the Kuopio Regional Institute of Occupational Health) and an educated guess was made that bioaerosol exposure could be at least one of the contributing factors. As we know, this guess hit upon quite the right thing. It was natural that the knowledge gained

47

N. Boschi (ed.), Education and Training in Indoor Air Sciences, 47–51.
© 1999 *Kluwer Academic Publishers. Printed in the Netherlands.*

in indoor air research was also utilized in training. Courses and other training forms were begun and extended at both universities, as will be described later. The HVAC laboratory started to arrange a national indoor air seminar already in 1983. It remained as a small workshop until 1987 when the Finish Society of HVAC Engineers took the arrangement responsibility and it was marketed also for engineers in industry. About 100 persons attended this seminar in 1987. After the Indoor Air =93 Conference in Helsinki, the awareness of the importance of good indoor air quality increased rapidly (the rapidly expanding moldy building problem was another reason; see the text below) and the the FISIAQ took now the organization responsibility. In this year, this seminar became a two day event with more than 800 participants.

In the 90's, a new culprit was found. It was the building industry which was responsible of many poor quality buildings erected during the years of the economic boom at the turn of the 90's. Most of these buildings suffered from moisture and mold problems, and the word mold building became almost a synonym for a problem building in Finland. As a consequence, the civil engineering departments of the technical universities and polytechnics joined the research, and, of course, these issues were also taken into consideration in their training programs. There was also an obvious need for continuous education for existing civil engineers in indoor air science; therefore, University of Kuopio coordinated a long-term indoor air specialist training program for unemployed civil engineers in 1997 (there were plenty of them available after a recent economic recession). Currently, this program has been renewed for employed civil engineers. All major research institutes in indoor air science have participated in arranging of these courses. Most recently, the Ministry of Education has provided funding for the Graduate School of Building Physics. It is coordinated by the Civil Engineering Department of the HUT. Other participating organizations include the Civil Engineering Department of the Tampere University of Technology, University of Kuopio, VTT, and NPHI. Because the frequency of moisture and mold problems was high in the building stock the amount of the building occupants having symptoms was also large; therefore, the above mentioned indoor air clinics were established.

Today, indoor air training and education is arranged by a number of organizations including many concerned in public or respiratory health. In the following, only four forms of training will be described:

1. The National Indoor Air Seminar,
2. The indoor air science program of the University of Kuopio,
3. The indoor air specialist training, and 4. The Graduate School of Building Physics.

3. National Indoor Air Seminar

This seminar has grown from a modest workshop of some twenty enthusiastic researchers in early 80's to today=s two-day mass event gaining also a lot of interest in public media. The seminar has now clearly adapted a double role; it is no more only a scientific conference but also an importants means to transfer knowledge. This year, the first day was ordinary scientific seminar even though there were already many Alaymen@ (civil and ventilation engineers, medical doctors, health inspectors etc.) among the audience in addition to indoor air researchers. The second day concentrated to explain the knowledge gained during the last year to the general public. In addition, an indoor air exhibition was arranged for ventilation companies, consults, book publishers etc. to present their products and services. Free entry has been one important factor contributing to the success of this seminar. The Ministry of Environment has provided the major financing for the seminar.

4. The indoor air science program of University of Kuopio

University of Kuopio has BSc, MSc and PhD programs in environmental science. Studies involve no charges;therefore, most students continue their studies until MSc. In order to achieve MSc the student has to gather 160 credits, 74 of them must be in environmental science. The recommended minors include chemistry, microbiology, and physics. The courses offered in environmental science are divided into modules of 10-20 credits, one basic studies module, four subject studies modules, and four advanced studies modules. Indoor air science studies do not constitute a separate module but are divided into several modules. The basic studies module includes the course in noise, illumination and thermal comfort. All seven courses of the exposure assessment module (subject studies level) are at least partly associated with indoor air (especially, indoor air hygiene I, chemical occupational hygiene I, indoor air measurements I, and radiation hygiene). The environmental health module (subject studies level) includes courses eg. in noise control, epidemiology, toxicology, and risk assessment. The advanced indoor air science courses belong mainly to the modules: physical environmental science and the adverse health effects of air impurities. The former includes an advanced noise course, nonionizing radiation, and advanced illumination and thermal comfort course. The latter consists entirely of indoor air courses, namely, chemical occupational hygiene II, indoor air hygiene II, indoor air measurements II, and ventilation. Finnish is the normal language but some courses (eg. Indoor air hygiene I) are held in English if there are any foreign students present. On the average, some 4-7 students prepare their MSc thesis on indoor air issues annually.

The topics of the doctoral training courses vary annually. Most common are courses in advanced epidemiology, biostatistics, environmental toxicology, and risk assessment. This year, several courses in exposure assessment and airborne allergens have been arranged. Many of the doctoral courses are held in English. Six students will defend their thesis in indoor air science this year. In addition, a few students have completed their doctoral studies in the medical faculty with a thesis dealing with building-related illnesses (less than one annually).

Valamo Conference on Environmental Health and Risk Assessment organized by Finnish Research Prgram on Environmental Health (a special four-year research program funded mainly by the Academy of Finland), University of Kuopio, National Public Health Institute and Kuopio Regional Institute of Occupational Health is a Gordon Conference type workshop to provide a forum for fruitful discussions on research ideas, disseminating education, and to create contacts between established scientists, post docs, and doctoral students. The conference venue is Valamo Orthorox Monastery whic provides an interesting atmosphere for the event.The conference language is English and foreign participants are welcome. The organizers cover the conference fee and accomodation in the monastry guesthouse. Travel expenses will not be subsidized. The first Valamo Conference was arranged last year. Then the idea was to gather together indoor air and ambient air researchers to facilitate collaboration between these two closely related research fields which have, however, quite different research practices for historical reasons. This year, the main themes will be epidemiology, indoor air, endocrine disruption, and social aspects in exposure.

5. Indoor air specialist training

The program is intended to train civil engineers to investigate and solve indoor air problems. The courses consist of lectures (400 h), exercises (20 h), and thesis work (400 h), together 30 credits. The main topics are: building microbiology, building-related illnesses, building physics and engineering, ventilation and air conditioning, dusts, odors, volatile organic compounds, and radon, juridical questions, project and total quality management, business administration, and building investigations.

Special attention is paid to behavior of building moisture, mold growth, building investigation methods, legislation, health effects, radon mitigation methods, building materials with low emissions, and maintenance of ventilation systems. Best specialists are used as teachers.

During the thesis period (10 credits), the students perform thorough building investigations and draw up plans for repair. They may also concentrate on ventilation problems or on some other relevant issue.

6. Graduate School on Building Physics

The Finnish graduate school system provides four year research assistant positions for doctoral students. They are assumed to complete their studies within this period of time. In addition to a financied possibility to concentrate to doctoral thesis work, the schools also arrange free doctoral courses for their students (other students may also participate). The speciality of the building physics school is its multidisciplinarity; its organizers include State Research Centre (material emissions), Departments of Chemistry (physical chemistry) and Machine Engineering (HVAC) of HUT, and National Public Health Institute and University of Kuopio (indoor air science) in addition to the civil engineering departments of the Finnish technical universities (Helsinki and Tampere). Thus, the courses cover similar wide range of topics as the indoor air specialist training. To some extent, the courses of these two training programs are interchangeable. In addition to lecture courses, seminars, excercises, and small research projects (called miniprojects) are arranged. Most students have civil engineering background but school has also students with MSc in environmental science, chemical engineering, and ventilation engineering.

7. Conclusions

Indoor air problems have gained considerable general interest in Finland and the need for indoor air science education has become obvious for many professionals, such as civil engineers, ventilation engineers, medical doctors, and health and building inspectors. In addition, these is a need of specialists in this field. It has been genarally understood that a multidisciplinary approach is required in this education, and several training organization networks have been established to achieve this. These efforts have already resulted in better building practices. More reliable building investigations have also become possible.

GRADUATE EDUCATION AND TRAINING IN INDOOR AIR SCIENCE - A CANADIAN APPROACH

F. HAGHIGHAT
Department of Building, Civil and Environmental Engineering
1455 de Maisonneuve Blvd. West
Concordia University, Montreal, CanadaH3G 1M8

Abstract:

In the past half century indoor air quality has become a major health, economy and comfort issue. Growing public awareness of this issue, coupled with the complexity and multi-disciplinary nature of the problems, has created a need to depart from the traditional education provided to engineers who become responsible to design and operate buildings. The range of training provided to building engineers must expand greatly as the process of designing, constructing and operating of building become more complex.

This paper first gives a historical overview of the Building Engineering Programme at Concordia University. It then focuses on its Graduate Programme, more specifically the Graduate Certificate Programme with the goal of evaluating the compatibility of this program for the International Education Program in Indoor Air Sciences.

1. Introduction:

Canada spreads between the latitude of 45 °N to the North Pole, and is composed of six climatic regions: Arctic, Northern Pacific, Cordillera, Prairie, and Southeastern. Consequently, the climate varies from high temperature s in the 30's °C in the summer to severe winters of well below -50 °C. As an example Montreal is located in latitude of 45 °N and located along the border of the Northern and Southeastern limits. It has dry-bulb temperatures of 30 °C or over, which is exceeded for 2.5% of the hours in July, while in January, it has dry bulb temperatures of -25 °C or lower, which is exceeded for 2.5% of the hours in that month. Therefore, designing a building envelope to withstand such a harsh climatical conditions while maintaining acceptable indoor environment and minimum energy consumption is a real engineering challenge. In Canada, twenty-five years ago, engineers performed this task with training in fields such as Mechanical Engineering, Chemical Engineering or Civil Engineering, not having knowledge of building science and/or formal background for such work. They mainly learnt this skill on site rather than at the university. For instance, an architect usually designed the building envelope, while its HVAC system designed was separately by an engineer. Subsequently, the dynamic thermal interaction between the different components was largely ignored.

N. Boschi (ed.), Education and Training in Indoor Air Sciences, 53–68.

The construction industry was booming in early 1970. Buildings were constructed with a sealed outer shell, so that all aspects of the environment within could be placed totally under human control. The specifications for the temperature, humidity, air movement, lighting, acoustics, etc. of this indoor environment, based on investigations of human comfort, was made in the design of these buildings before construction to provide an acceptable environment for the occupants. Since then, there has been growing concern and uncertainty with the quality of the indoor environment due to commonly attributed adverse effects on comfort and health. This uncertainty has resulted from a number of factors: changing design and operation of buildings to reduce energy consumption, tightening of the building envelope to reduce uncontrolled air leakage which contributes to the moisture deterioration of the building envelope, new materials and related emission, and the general increased public awareness of the relationship between health, comfort and productivity and the environment.

This suggests an interdisciplinary approach to the building process and, the requirement for a new *educational paradigm*, to address the planning, design, construction, operation and maintenance of healthy, comfortable and energy efficient buildings.

With this objective, the Centre for Building Studies (CBS) was established at Concordia University in 1977. The program focuses its attention to interdisciplinary areas of building, including:

Building Envelope which focuses on the analysis and design of building envelopes, including durability, heat and moisture transfer and interaction with the indoor environment.

Building Science, which focuses on the analysis and control of the physical phenomena affecting the performance of building materials and building enclosure systems.

Construction Management which covers construction techniques, construction processes, planning, scheduling, projects tracking and control, labour and industrial relations, and legal issues in construction.

Energy Efficiency, which deals with the analysis, design and control of energy-efficient building and HVAC systems, solar energy utilization and intelligent buildings.

Indoor Environment covers the environmental aspects in the design, analysis and operation of energy-efficient, healthy and comfortable buildings

2. Graduate Programme

The Department of Building, Civil and Environmental Engineering (formally known as CBS) offers a wide range of graduate programs in Building and in Civil Engineering. The Department houses the Centre for Building Studies, an interdisciplinary research centre with international recognition that played a key role in the development of the Building Engineering discipline in Canada.

The focus of Building Engineering is to advance the body of technical knowledge in planning and design of built facilities, their construction processes, their operation and maintenance, and their interaction with, and impact on, the surrounding environment. This programme leads to Graduate Certificate in Building Studies, Master of Magisteriate in Applied Science (M.A.Sc.) in Building Engineering, Master of Magisteriate in Engineering (M.Eng.) (Building), and Ph.D. in Building Studies.

The Department offers two 45-credit programs leading to the *M.A.Sc. or M.Eng.* degrees with specialization in one of the following four main subjects areas: Building Sciences, Building Environment, Building Structure and Construction Management. The Department also offers a Ph.D. Program, which is designed to provide students an opportunity to obtain the greatest possible expertise in Building Studies through a combination of courses and intensive research.

The *Graduate Certificate* in Building Studies is for practising engineers, architects, educators and researchers whose professional activities encourage them to improve their expertise in one the following sub-disciplines within Building Engineering: Building Science, Building Envelope, Construction Management, Energy Efficiency, Indoor Environment. A fully qualified candidate is required to complete a minimum of 15 credit. Nine credits must be taken of core courses in the area of concentration and the balance of 6 credits may be taken from the other courses offered by the department [1].

The following courses are offered within these sub-disciplines:

Building Science
Core Courses: Building Science (BLDG 661), Modern Building Material (BLDG 662), Indoor Air Quality (BLDG 675). Electives: Fire and Smoke Control in Buildings (BLDG 665), Building Acoustics (BLDG 672), Building Illumination (BLDG 673), Principle of Solar Engineering (ENGR 660), Solar Energy Material Science (ENGR 666).

Building Envelope
Core Courses: Building Enclosure (BLDG 660), Building Science (BLDG 661), Thermal Performance of the Building Envelope (BLDG 666), Electives: Structural Systems for Buildings (BLDG 606), Wind Engineering and Building Aerodynamics (BLDG 607), Computer-Aided Building Design (BLDG 659), Modern Building Materials (BLDG 662), Building Illumination (BLDG 673).

Construction Management
Core Courses: Building Economics 1 (BLDG 656), Project Management (BLDG 657), and Construction Processes (BLDG 683). Electives: Decision Analysis (BLDG 658), Construction Planning and Control 1 (BLDG 680), Labour and Industrial Relations in

Construction (BLDG 681), Legal Issues in Construction (BLDG 682), Construction Planning and Control ll (BLDG 684), Project and Cost Estimating (BLDG 685).

Energy Efficiency

Core Courses: Thermal Performance of the Building Envelope (BLDG 666), Building and Environment (BLDG 670), Mechanical Systems in Building (BLDG 671). Electives: Building Science (BLDG 661), Automatic Controls for Building Environmental Systems (BLDG 674), Intelligent Buildings (BLDG 676), Passive Solar Heating (ENGR 665), and Principles of Solar Engineering (ENGR 660).

Indoor Environment

Core Courses: Building and Environment (BLDG 670), Building Illumination (BLDG 673) and Indoor Air Quality (BLDG 675). Electives: Computer-Aided Building Operation (BLDG 611), Thermal Performance of the Building Envelope (BLDG 666), Building Acoustics (BLDG 672), and Principles of Solar Engineering (ENGR 660).

Students from almost all the departments in the faculty of engineering (mechanical, civil, chemical), science (chemistry, physics, and biology), Arts (architects) have attended the Graduate Certificate program. For this reason, an attempt has been made to avoid the pre requisites as far as possible. However, when a student with no background in engineering wants to follow the program, they are encouraged to get the approval of the instructor. In some cases, the student may have to take some additional courses. Experience with this program has generally been quite favourable and the indications are that the package of courses offered in this sub-disciplines is reasonably successful in exposing students to the related discipline and application.

Three core courses, BLDG 670, BLDG 673 and BLDG 675, are suggested for students seeking to develop expertise in the area of Indoor Environment. BLDG 670 focuses on requirement for human comfort, primarily thermal, acoustical and visual. BLDG 673 covers the calculation techniques and qualitative aspects of good luminous design. BLDG 675 deals with the effects of contaminants on health and well being of occupants, and investigates the causes and methods of alleviating and curing unhealthy environments. In BLDG 675 course, through discussion and by means of guest lectures, some aspects of human physiology, biology are also presented to the students. As part of the requirement for this course each student must find and investigate a building following the three-part approach to investigation: walkthrough of occupied spaces, ventilation system inspection, and measurements and air samples. At the end of the term each student presents his/her project.

Students, who follow the Indoor Environment sub-discipline, have a good knowledge of the source of pollutants, HVAC/ventilation, source control, monitoring, diagnostics, modelling, planning, management, design, etc. Therefore, they can systematically investigate the building as a *building engineer* and propose engineering solutions.

3. Indoor Air Sciences and the New Philosophy

Extensive indoor air investigations around the world have indicated that there is more in investigating a building, than just measuring physical and chemical indicators. Haghighat et al. [2] examined the relationships between the indoor environment parameters on two floors of an eleven-story building, as perceived by the occupants and as measured objectively. They showed that complaints reported by the occupants were associated with perceived rather than measured levels of indoors environmental parameters. The study was conducted over a 4-week period and consisted of measuring environmental parameters, and of administering a questionnaire on comfort and health, to 450 occupants. Most noteworthy in the responses was that more than 34% of the occupants expressed that the air was dry. The measured relative humidity ranged from 40 to 65%. More than 32% of the occupants expressed that in general, the thermal environment was unsatisfactory, even though almost all the measured thermal comfort parameters complied with the ASHRAE comfort standard. ASHRAE defines an acceptable thermal environment as "an environment that at least 80% of the occupants would find thermally acceptable" [3].

The ANSI/ASHRAE Standard 55-92 "Thermal environmental conditions for human occupancy" [3] is used extensively in Canada, as a reference for comfort levels. As more and more studies of Canadian buildings in the cold climate are emerging, it is apparent that the measured parameters satisfy the comfort limits as set out by ASHRAE, yet it is found that less than 80% of the occupants are satisfied [4]. ANSI/ASHRAE Standard 55-92 is based almost entirely on data from climate chamber studies performed in temperate climates. This perhaps explains the discrepancies between occupant satisfaction in a cold climate and satisfaction of workers in a temperate climate.

Building occupants often react to their environment in markedly different ways, and it is often difficult to identify the sources, which are the cause of particular problems. The symptoms observed in building occupants are varied, and they depend greatly on the thermal parameters (air and wall surface temperatures, air velocity and fluctuation, relative humidity, clothing, etc.) contaminants (type and concentration), lighting, psychosocial factors (office space, personal control, job satisfaction, relation with co-worker, etc). The symptoms include headaches, dizziness, cough, eye irritation, unpleasant odours, fatigue, respiratory problems, and nose and throat irritation: these are so called the symptoms of the "Sick Building Syndrome" (SBS).

Therefore, symptoms of SBS are of multifactorial origin. It has been found that psychosocial work characteristics, such as workload and job satisfaction, as well as worry and reorganisation, are factors that have a significant impact on the risk of developing the symptoms of "sick building syndrome" (SBS)[5,6]. Skov et al. [7] and Eriksson et al. [5] found a strong relationship between "satisfaction with superiors" and the prevalence of mucosal and general symptoms. Relations to supervisors were associated with a slightly increased risk. Skov et al. [7] reported a strong association between "satisfaction with colleagues" and symptoms. However, Eriksson et al. [5] did not find such an association, nor did a feeling of poor work status have any impact on the risk of having symptoms. Eriksson et al. [5] surveyed almost 6000 office workers from three cities in Sweden. They found that poor workplace satisfaction (salary, benefits, opportunities for growth, and personal development) was associated with a significantly higher risk (also supported by Zweers [8]). However, Hedge [9] did not find any association between job satisfaction and SBS. Eriksson et al. [5] found that satisfaction with work seems to be irrelevant in this context, whereas satisfaction that refers to the workplace, i.e. rewards and opportunities for growth, appears to be a significant factor. They reported that an adverse psychosocial work environment might constitute a stress that causes symptoms through psychophysiological reactions. Thus, SBS may be regarded as a psychosomatic disorder, with psychological distress being expressed through physical symptoms. They also suggested that an adverse psychosocial environment may make the individual more attentive to discomfort and health, and to potential causes in the physical environment, and this could affect reporting behaviour. Another suggestion was that discomfort with the physical environment affects perception of the psychosocial environment. Or, a poor psychosocial environment may make the individual more susceptible to different adverse indoor climate factors.

Haghighat and Donnini [4] reported the results of a detailed measurement in twelve mechanically ventilated buildings. The indoor air quality, thermal comfort, energy consumption, and perception of occupants were measured in these buildings. A total of 877 subjects participated in the questionnaire survey. The questions included in the questionnaire dealt with health, environmental sensitivity, work area satisfaction, personal control of the workstation's environment, and job satisfaction. The occupants assessed their health status by rating ten symptoms. The most frequently occurring symptom was "fatigue", in both seasons; and the least occurring symptom was "dizziness", in both seasons, Figure 1 and 2. Figures 3 and 4 show the level of occupants' satisfaction with fifteen aspects of their job. This study also showed that workers report more symptoms when they perceived IAQ to be poor. The more dissatisfied with the IAQ was the occupant, the more often the occupant self-reported health symptoms.

These results indicate that a multi-disciplinary approach is required involving the engineering, social and medical communities working towards a common mission to educate the *future indoor air science engineer*. Then we can see that the new and existing buildings meet the needs of the future. It is only through a better understanding of the relationship between physical attribute of the indoor environment and occupant health, satisfaction(s) and well being, that the engineering scientists can develop economically feasible approaches for improving existing buildings, and disgning healthy, comfortable and energy efficient new buildings.

The new discipline of Indoor Air Science should address the Building and Occupants as an integrated system. Consequently, the players have to come together from various disciplines to focus their attention in areas such as: the health and physiological aspects of the indoor environment (temperature, relative humidity, carbon dioxide concentration, etc.), the effects of specific environments on human comfort and health (VOC concentration).

The main objective of the new program must be to introduce all students to the *real* interdisciplinary aspect of the indoor air environment. This includes providing students with sufficient theoretical background in engineering, psychology, biology, economy, chemistry to grasp the general concepts, as well as some practical experience on investigation techniques including measurements and monitoring.

4. Conclusion

We live in a rapidly changing world, It is the responsibility of educational institute to prepare graduates to integrate well into the work place and be able to adapt to the changes in technology that they will encounter almost from the time they graduate.

The Graduate Certificate program at the Department of Building, Civil and Environmental Engineering at Concordia University has been very successful providing the students with sufficient background in Building Engineering. However, some of the aspects of human physiology, psychology, biology, and basic chemistry are not covered in the suggested courses for the Indoor Environment option. Many "real life" building investigations are multidisciplinary in scope. Feed back received from students indicates that this should be improved.

Therefore, all points of view indicate that the graduate certificate program is an excellent format for the proposed Indoor Air Science program. It can be modified with some minor modification and adopted it for the students with different background.

60

5. References

1   Concordia University Graduate Calendar, 1998-1999, Concordia University, Montreal, Canada

2   Haghighat, F., Donnini, G., and D'Addaria, R. (1992) Relationship Between Occupant Discomfort as Perceived and as Measured Objectively, *Indoor Environment* 2, 163-174.

3   ASHRAE 1992, ANSI/ASHRAE Standard 55-1992, Thermal Environmental Conditions for Human Occupancy. Atlanta: American Society of Heating, Refrigerating, and Air Conditioning Engineers, Inc. USA.

4   Haghighat, F., Donnini, G. (1998) Impact of psycho-social factors on perception of the indoor air environment studies in 12 office buildings, *Building & Environment*.

5   Eriksson, N. Hoog, J., Stenberg, B., Sundell, J. (1996) Psychosocial factors and the sick building-syndrome - a case-referent study, *Indoor Air* 101-110.

6   Karasek, R. A. (1990) "Lower health risk with increased job control among white-collar workers", *J. of Organization Behaviour* 11, 171-185.

7   Skov. P., Valbjorn, O., Pedersen, B.V. (1989) Influence of personal characteristics, job-related factors and psychosocial factors on the sick building syndrome, *Scandinavian J. of Work, Environment and Health* 15, 286-295.

8   Zweers, T., Preller, L., Brunekreef, B. and Boleij, J.S.M. (1992) Health and indoor climate complaints of 7043 office workers in 61 buildings in the Netherlands, *Indoor Air*, 2, 127-136.

9   Hedge, A. (1988) Job stress, job satisfaction, and work-related illness in offices, *Proc. of the 32nd Annual Meeting, Human Factors Society*, Santa Monica, CA: Human Factor Society, Vol. 2, PP. 777-779.

Caption

Figure 1: Self-reports of health symptom frequency (summer)

Figure 2: Self-reports of health symptom frequency (winter)

Figure 3: Job satisfaction ratings (summer)

Figure 4: job satisfaction ratings (winter)

61

Figure 1-a: Self-reports of health symptom frequency (summer)

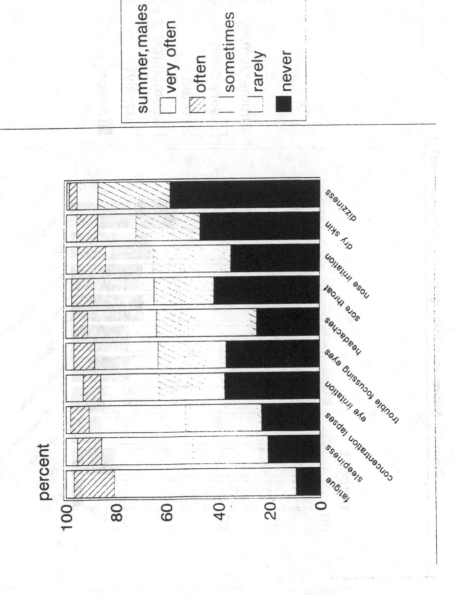

Figure 1-b: Self-reports of health symptom frequency (summer)

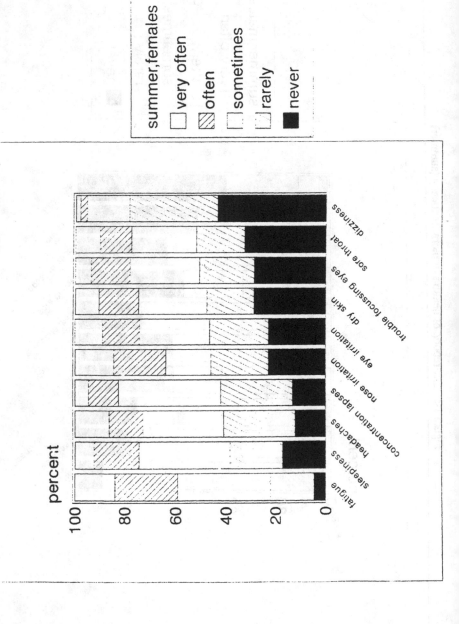

Figure 2-a: Self-reports of health symptom frequency (winter)

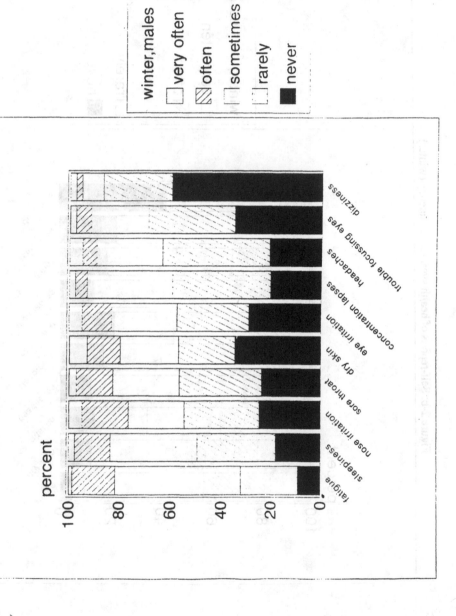

64

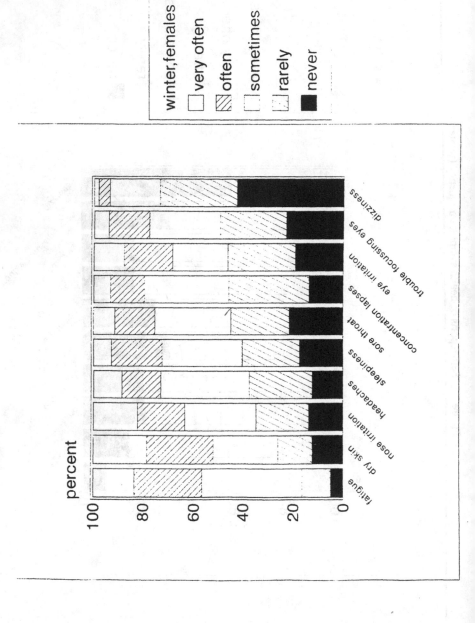

Figure 2-b: Self-reports of health symptom frequency (winter)

65

Figure 3-a: Job satisfaction ratings (summer)

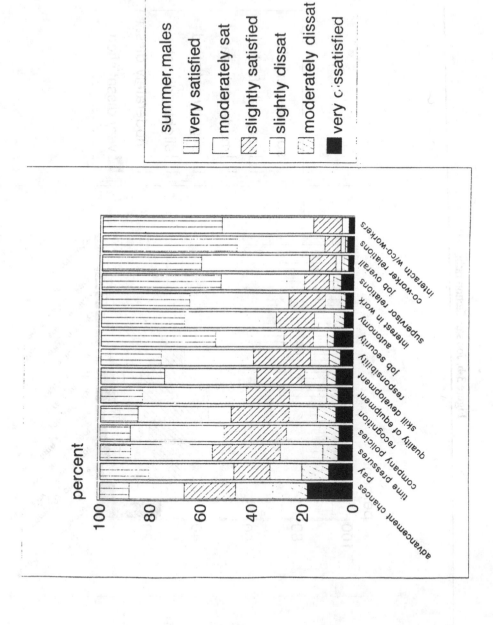

66

Figure 3-b: Job satisfaction ratings (summer)

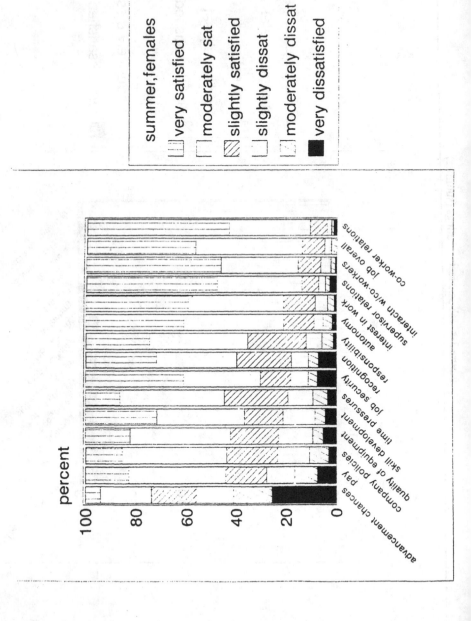

summer,females

- very satisfied
- moderately sat
- slightly satisfied
- slightly dissat
- moderately dissat
- very dissatisfied

Figure 4-a: Job satisfaction ratings (winter)

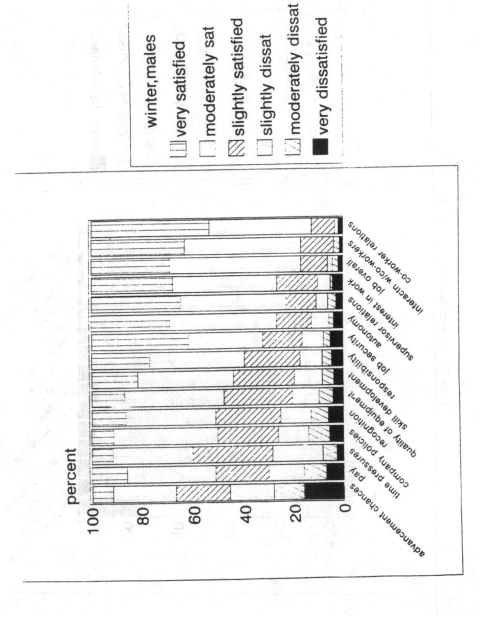

Figure 4-b: Job satisfaction ratings (winter)

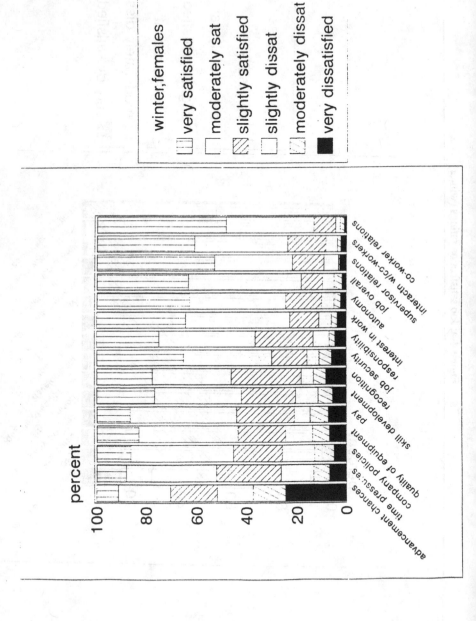

INDOOR AIR FUNDAMENTALS AND GRADUATE EDUCATION IN THE CZECH REPUBLIC

V. BENCKO, I. HOLCÁTOVÁ

Institute of Hygiene and Epidemiology, First Faculty of Medicine,
Charles University, Prague, Czech Republic

1. Abstract

With the teaching schedules in indoor environment related health aspects for the pregraduate students we have, al least, scored a relative success at the Faculties of Medicine and Sciences of our Charles University and Faculty of Architecture at the Czech Technical University in Prague. Another positive move is our share in the postgraduate training of our hygienists, physicians engaged in public health and clinicians like allergologists, immunologists and paediatricians. Though our achievements in the education sphere should not be overrated they nevertheless appear promising for the future.

2. Introduction

In our country was experience with indoor pollutants, especially with asbestos and formaldehyde and workers from technical universities had adequate experience on indoor humidity and temperature related problems. Medical doctors were interested in those technical parameters especially from the point of view of occupational medicine. So we have not started from the point zero [1].

After the political changes in 1989 students at our faculty of medicine preferred to restrict all non-clinical courses. So we had not a simple task - to change our courses into more attractive way for future "clinical doctors". The main problems were the technical topics, which are some times difficult to understand and more over students believed that they would never need such a knowledge. From that point of view the problems of indoor environment, sick building syndrome and buildings related diseases, especially in the context of allergies and hypersensitivity symptoms related to buildings seemed to us very convenient subjects.

3. Implementation of indoor fundamentals in graduate education

We have started our educational efforts by one-hour seminar. Our basic knowledge came from the proceedings of the NATO/CCMS conference in Erice [2] and from the

N. Boschi (ed.), Education and Training in Indoor Air Sciences, 69–74.

proceedings of the Indoor Air '90 in Canada [3] and of course, from the previous studies of our specialists concerning formaldehyde, radon and indoor related health problems in occupational settings [4 - 9]).

At the beginning this topic was not too much interesting for our students. Then we have got more sources of information - journals Indoor Air and Indoor Environment, proceedings from other conferences and books [10 - 12].

We presented more clinical problems, which are connected with the indoor environment and of course we have used results from our clinical studies. All these changes attracted an interest of students by presenting them the practical solutions of some indoor air problems with practical health implications.

Our courses on hygiene and epidemiology at the First Faculty of Medicine, Charles University of Prague, are scheduled at the fifth year of medical study. We try to attract students who are more interested in clinical subjects, to some health related indoor environmental topics or better to say, let them understand the necessity of some basic knowledge, "fundamentals" from these topics even for clinicians.

Now the seminars on health aspects of air pollution lasts about five hours. One third of the allotted time we spend on outdoor air pollution and most of the time is spent on indoor related problems. At the end of every block of seminars we asked students to evaluate the level of course (they spend at our institute three weeks block course on hygiene and epidemiology). Air pollution and sick building syndrome - this is the title of the seminary - have got every time one of the best evaluations and we can spend on that topic even more time because of the students interest. We have published recently new textbook on hygiene [13] where the chapter Indoor Environment is rather extensive and additional technical aspects are discussed in some other chapters (noise, lighting, radon, and electromagnetic fields).

Besides of teaching of our students of medicine we offer only slightly modified course on hygiene to students of the Faculty of Sciences on using the same teaching schedule and teaching textbook on hygiene. This course on hygiene is obligatory for students of subject Environmental Science. When teaching the subject indoor environment related problems in this case we just put less attention to health specific details.

We try to infiltrate to the technical university and we are trying to extend contacts with architects, because we feel this collaboration be very useful in prevention of indoor environment problems.

First real steps we made five years ago, when we started to practice to give two 2-hour seminars concerning health aspects of indoor environment related subject for students of architecture within subject Fundamentals of Building Design. Now there is under preparation a new textbook on this subject in which a special chapter will be devoted to health aspects of indoor environment related problems.

In teaching this kind of students we try to demonstrate not only "frightening examples" of indoor air history like legionnaires pneumonia, asbestos or radon carcinogenicity, but as well at least some positive cases like potential beneficial influence of electroionizing microclimate and influence of different kinds of building materials in that context [14].

4. Postgraduate education

So we are quite successful in attracting the young generation of our students and future physicians to the indoor environment problems. But we need more contacts with the clinicians. In that field we succeeded only in a limited extent. The first contacts we have had with specialists in ENT (ear, nose and throat), but the best collaboration we have with alergollogists and paediatricians. They are interested in that topic, because they use this knowledge in treatment of their patients more frequently than other clinicians. In 1992 we started to discuss with clinicians some special problems concerning the indoor environment and in 1993 we have got the grant on that topic, so we could start at least some limited research concerning asbestos, formaldehyde and humidity [15 – 23].

We are giving lectures in postgraduate courses for allergologists [24], immunologists, paediatricians and even hospital managers and every time these lectures are quite successful with interesting discussion. Many participants tell us that they never suppose such common situation as e.g. dry indoor air during the winter could be so important from the point of view of health problems of their patients.

Last but one year our Institute for Postgraduate Education of Health Professionals in Prague has organised for the first time a special course on indoor air topics for hygienists and there were quite a lot of colleagues who were interested in it. Than the last year was organised one-day course on indoor environment related health problems and such a courses are to be organised since at least once a year.

We are also regularly invited to have lectures on seminars and courses organised by the Czech Society of Environment Techniques and we have published pertinent papers in local professional journal e.g. ARCHITEKT (Architect), STAVEBNÍ ROZHLEDY (Prospects of. Building Construction).

We do hope as well to continue the postgraduate courses for some other medical specialists in the future.

5. Discussion

Environmental pollution is one of the more pressing problems of our age. Air pollution is not a new problem. For many years researchers have been interested in sources and health effects of air pollution. In our country there were a lot of studies concerning air pollution and health status especially of the children. Nevertheless it was difficult or some times rather impossible to publish such a complex studies on environmental pollution including indoor air related problems together with health status related data.

After the 1989 political changes in our country people believed to get all, not restricted information about the effect of environmental pollution to their health as in these years in many journals there was published a lot of information about bad health status of our population in heavily polluted regions in previous Czechoslovakia.

Unfortunately, no such a strict evidence or narrow connection between the health status of population and outdoor pollution were demonstrated with exception of above mentioned heavily polluted regions. From epidemiological studies seems to be the lifestyle more important as environmental pollution.

Hygienists were the first physicians who were attracted to this topic, as their field of work is more complex, directed not only to the medical point of view but also to neghligible extent from technical one as well. They used to check and evaluate all kinds of projects so they need to understand the both sides of this subject.

6. Conclusion

The best proposal how to continue in our efforts not only in our country but probably for all the CEE countries seems to be the collaboration inside the countries, exchange of knowledge and also international co-operation among the countries as could be seen in NATO/CCMS Pilot Study on Indoor Air Quality [25]. It is recommendable to join technical equipment and know-how in internationally based collaborative projects with more experienced and better-equipped research teams working in NATO countries [26].

We hope that we are quite successful in spreading the knowledge about the indoor environment problems in a community of medical doctors and students of sciences specialised in environmental sciences. We made also some progress in collaboration with the technical universities. But people from the practice still don't care too much about these problems. So now we try to spread the knowledge on that topic in a general population, trying to influence the attitude of other specialists as well. We try to use "public relations" - contacts with journalists, we try to attract their interest to this topic and we are partly successful.

7. References

1. Bencko V. (1994) Health Risk of Indoor Air Pollutants: A Central European Perspective. Indoor Environ., No.3, 213 - 223.

2. Maroni, M., Berry, M. A. (ed.) (1989) The Implications of Indoor Air Quality for Modern Society. Report on a Meeting Held in Erice, Italy. NATO/CCMS Report No 183, pp. 67 - 76.

73

3. Walkinshaw, D. (ed.) (1990) Indoor Air '90. Proceedings of the Fifth International Conference on Indoor Air Quality and Climate, Toronto, Canada, 5 volumes.

4. Hanzl, J., Rössner, P., Klementová, H. (1985) Cytogenetic Analysis in Workers Occupationally Exposed to Formaldehyde (in Czech) Cs. Hyg. 30, 7-8, 403-410

5. Dobiáš, L., Hanzl, J., Rössner, P., Janèa, L., Rulíšková, H., Andìlová, Š. Klementová, H. (1988) Evaluation of the Clastogenic Effect of Formaldehyde in Children in Kindergarten and School Facilities (in Czech) Cs.Hyg. 33, 10, 596-604

6. Chýlková,V., Turková, M., Procházková, J., Vo enílková, H., Tmìjová, M. (1990) Action of Formaldehyde and Toluene on Selected Indicators Humoral Immunity in Serum and Saliva of Apprentices (in Czech) Cs.Hyg., 35, 2, 76-86

7. Srb, V., Rössner, P., Zudová, Z., Pohlová, H., Koudela, K., Vo enílková, H., Tmìjová, M., Eminger, S., Kubzová, E., Vodvárka, Z., Zaydlar, K. (1990) Action of Formaldehyde and Toluene on Selected Cytogenetic Indicators in Apprentices (in Czech) Cs.Hyg.,35, 2, 66-75

8. Janèa, L., Dobiáš, L., Andìlová, Š. (1986) Formaldehyde in Indoor Air from the Genotoxicyty Point of View. (in Czech) AHEM, Suppl. 5, 47-53

9. Holcátová I., V. Bencko V. (1997) Health Aspects of Formaldehyde in the Indoor Environment. Czech and Slovak Experience. Centr eur J public Health, 5, No. 1, 38 - 42.

10. Samet, J. M., Spengler J.D. (ed.) (1991) Indoor Air. A Health Perspective. The John Hopkins University Press.

11. Bieva, C.J., Courtois, Y., Govaerts, M. (eds.) (1989) Present and Future of Indoor Air Quality. Proceedings of the Brussels Conference, Excerpta Medica.

12. Jokl, M.(1989) Microenvironments: The Theory and Stress of Indoor Climate.Thomas, Springfield, III.

13. Holcátová I., Bencko V. (1998) Indoor Environment (in Czech), in Bencko V. (ed.) Selected Chapters from Hygiene. The Textbook for Medical Students. 1st Medical Faculty, Charles University, Prague, pp. 39 – 46

14. Lajèíková, A., Mathauserová, Z., Bencko, V. (1999) Electro-ionic Microclimate and Materials Used in the Indoor Environment. Centr eur J public Hlth, 1 (in press).

15. Holcátová, I., Holcát, M. (1994) Indoor Air Quality and Respiratory Diseases. In: L. Bánhidi, I. Farkas, Z. Magyar, P. Rudnai (eds.), proc. Healthy Buildings '94,: B-udapest, pp. 683 –68

74

16. Holcátová, I., Holcát, M. (1995) Indoor environment and upper respiratory airways disorders (in Czech), in proc. *Living conditions and health*, Bojnice, Vol. II., pp. 307 - 312.

17. Holcátová, I., Gajdoš, P. (1995) Evaluation of Indoor Environment in Allergy Children, in G. Flatheim, K. R. Berg, K. I. Edvardsen (eds.), proc. *Indoor Air Quality in Practice - Moisture and Cold Climate Solutions*. Oslo, pp. 266 - 277.

18. Holcátová, I., Rybníèek, O. (1995) Aerobiology of indoor environment, in Rybníèek, O. (ed.) *Annual report of the PIS 1994*, Brno, pp. 6 - 17.

19. Holcátová I., Holcát, M., Rameš, J. (1995) Effect of dry air during the heating period on people with chronic respiratory diseases. in M. Maroni (ed.) Proc. *Healthy Buildings '95*, , vol. 1, University of Milan, pp. 443-448.

20. Holcátová I. (1995) Thermal and humidity comfort or heating stress in hospitals? in M. Maroni (ed.) Proc. *Healthy Buildings '95*, vol. 2, University of Milan, pp. 1077 - 1080.

21. Holcátová I., Gajdoš P. (1995) Concentration of volatile organic compounds and formaldehyde in buildings in connection to health status and quality of life of the dwellers, in Knight J.J., Perry R. (eds), Proc. *Volatile organic compounds in the environment*, London, pp. 253 -260.

22. Holcátová, I., Holcát, M. (1995) Indoor Environment and Quality of Life, in D. Petráš (ed.), *Proc. Indoor Climate of Buildings - Health and Comfort vs Energy*, SSTP, pp. 27 - 31.

23. Holcátová I. (1995) Possible changes of indoor environment and its effect on chronic diseases of the respiratory airways (in Czech). PhD. Prague.

24. Holcátová I. (1995) Outdoor and Indoor Air Pollution (in Czech), in Špièák V. (ed.) *The Textbook for the Postgraduate Educational Courses for Allergologists, Paediatricians and Pneumologists*. Mediforum.

25. Maroni, M., Axelrad, R., Tabunschikov, Y.A. (eds.) (1997) Pilot Study on Indoor Air Quality, Phase II. NATO/CCMS.

26. Holcátová I. (1999) Health Effects of Indoor Air Pollutants – Difficulties of Evaluation. Centr eur J Public Hlth, 1, (in press).

THE TECHING OF INDOOR AIR QUALITY AT THE DEPARTMENT OF MECHANICAL ENGINEERING AT THE TECHNICAL UNIVERSITY OF BUDAPEST

L. BÁNHIDI

Technical University of Budapest, Hungary

1. Abstract

At higher educational level indoor air quality is taught at the Department of Building Services Engineering at the Technical University of Budapest in Hungary. Its basic subject is "comfort theory", 60% of which deals with this subject. Other subjects dealing with indoor air in building services engineering studies (e.g. ventilation and air condition technologies) are based on these basic notions. Students also have to address the subject of indoor air quality in the majority of these final thesis projects.

2. Introduction

Mainly two branches of studies, medical research and technical research deal with indoor air quality in closed spaces and its impact on human beings. The former chiefly concentrates on the impact on humans and diagnostic problems, while the latter examines the technical possibilities and conditions of ensuring optimal or permissible air quality from the humans' point of view. The representatives of the two fields must work together in close co-operation: technical researchers have to meet the physiological and hygienic requirements of air quality and doctors have to be familiar with the technical possibilities and limitations of the specifications they have required and proposed. Feasibility is obviously restricted by the current economic situation, therefore it is necessary to know the technical designer the optimal and permissible values for indoor air quality.

Representatives of the technical field must be familiar with the subject at higher educational level. In Hungary it is taught at the Technical University of Budapest. Theoretically this knowledge is mainly necessary for students of the departments of architecture and mechanical engineering for a different purpose, however and to a different extent. For instance the architects have to be duly careful when choosing building materials to avoid sources of harmful pollutants. The job of mechanical engineers (including building services engineers) is to ensure the adequate air quality in closed spaces by removing pollutants, avoiding outdoor air pollution mainly with the help of (natural and/or mechanical) ventilation.

At the Department of Architecture of the Technical University of Budapest this subject is taught for one semester in the framework of postgraduate studies, while at the

75

N. Boschi (ed.), Education and Training in Indoor Air Sciences, 75–80.

Department of Mechanical Engineering it is part of the regular instruction. We would like to present the latter in detail.

3. The structure of the curriculum for mechanical engineers

Mechanical engineering studies last for 9 semesters in 24 classes per week. In the first 6 semesters the basic subjects (mathematics, mechanics etc.) are taught. In the 7th semester students have to choose two so-called modules.

In the past few years our aim was to provide students with the choice of the widest ranging overview and instruction in their mechanical engineering studies. Therefore if a student opted for the building services module, he could not choose as his second optional class other related subjects such as energetics, hydrodynamic machinery, caloric machinery etc. Last year this restriction was lifted because it was generally thought illogical. The building service module is presented in Table 1 as the typical struction of all module. The teaching aim of this module can be summarized as follows. The aim of the module: the training of qualified engineers, who have a theoretical knowledge of planning heating, ventilating, air conditioning, water and drainage systems and parts of these systems and are capable of technically preparing their construction and managing them. In the framework of the module, students study the necessary technical, economic and environmental subjects. After graduation, they are able to process and use on their own the latest technical findings of building services engineering.

Table 1 can be explained as follows:

Within the module subjects are put into three groups:
- compulsory
- optional subjects, but it is compulsory to choose one from them
- optional subjects.

As it is seen, the structure is based on the so-called credit system where students have to obtain a given number of credits in each semester. In addition, in certain subjects there are no called
- "contact" classes which are the sum of a lecture, a seminar and the laboratory practice,
- home classes: the amount of work students have to put into a subject at home.-

V/f means that students will either have to take an exam or will get a mark on the basic of their activity during semester (from 5: excellent to 1: fail).

The first column of all three tables indicates the semester when the subject has to be taken. The number of classes indicates the compulsory number of classes per week for 14 weeks.

Compulsory subjects and the number of classes do not require any further explanation. Students have to choose classes per week from the compulsory subjects and classes per week from the optional subjects. Their choice may depend on the number of credits, their interests etc.

Optional classes comprise all subjects taught at the university (or maybe at other universities). This list only includes subjects closely related to the building services module.

At the oral defence of dissertations, students have to take an exam in three subjects. Choice of topics in the building services module:
- Comfort theory
- Heating techniques
- Air technique systems
- Air conditioning techniques
- Water supply, gas supply, drainage systems.

4. The teaching of indoor air quality

Comfort theory comprises the following main topics (listed with the number of classes per term):

- Thermal comfort	9 classes
- The basics of acoustics	3 classes
- Illumination and day light	3 classes
- Comfort simulation	3 classes
- Air quality	21 classes
- Written exam	3 classes

This is a total of 42 classes (3 classes per week for 14 weeks). The list shows the major role of indoor air quality.

TOPICS IN INDOOR AIR QUALITY

1. Indoor air quality
   1.1. Elements: air quality protection - health protection at work, MAK, MIK numbers, types of pollutants, sources of pollutant.
   1.2. Physics of gases: State indicators (t,p,v,p,u,i), gas laws (Boyle and Mariott, Gay-Lussac universal gas law), gas mixtures (p, m, R).
   1.3. Concentration of pollutants: units, models for the changes in concentration, a, source with permanent emissions, b, for the penetration of a given quantity of pollutants.
   1.4. Elements of physiology and hygiene: Pettenkoffer number, CO_2 concentration, breathing, the structure of lungs, oxygen uptake, O_2 concentration, respiratory air, fresh air demand, composition of dry air.
   1.5. Olfaction, olfactometric: mechanism of olfaction, olfactory receptors, olfactory epithelium, olfactometric, Weber-Fechner law.
   1.6. Air quality-FANGER's theory: Pettenkoffer, many pollutants (~8000 types). new evaluation of emissions (olf), new unit of air quality (decipol dp), dp

measurement-etalon, comfort equation of air quality, fresh air demand on the basis of air quality requirements, Low-olf building.

1.7. Pollutants in comfort space: CO_2, radon, asbestos, formaldehyde VOC, TVOC, tobacco smoke, air conditioning system.

1.8. Air quality dimensioning: international standards, recommendations (pr ENV 1752, DIN 1946/2, EU Project), V_{fresh} based on air quality comfort pr ENV 1752, V_{fresh} based on health aspects DIN 1946/2, new results of EU Project, fresh air demand (per capits, area, air quality requirement), degree of contamination, efficiency of ventilation-ventilation (mixing ventilation, displacement ventilation, Quellüftung), air change - pollutant change, nominal time constant of room, local time constant.

1.9. Measurement of air quality: Fanger's measuring method, objective measuring methods, (continuous sampling); evaluation of air quality with trace gas.

1.10. Sick Building Syndrome: origin of SBS problems, SBS problems and reasons.

2. Air filtering

2.1. Dust contamination of outdoor air: dust concentration, emission values in Hungary, particle size distribution in air.

2.2. Origin and characteristics of dusts: natural and artificial dusts, physiological impact of dusts Johannesburg distribution, classification, measurement and precipitation in function of diameter.

2.3. Disperse systems.

2.4. Filtering theory.

2.5. Precipitation impacts in filters.

2.6. Pressure losses of fibre filters.

2.1. Filters.

2.8. Commercial forms of air-conditioner filters.

2.9. Filtering techniques in air-conditioners.

5. Summary

As the above shows, indoor air quality is taught to students of building services engineering at the Technical University of Budapest at an adequate level and in necessary detail. In additional, special strainings are held for post-graduate students. The available Hungarian and international bibliography is listed in References.

6. References

Bánhidi, L. (1993) Joint sick building studies in Hungary with Yu-GRD participation. Research and Healthy Buildings, CIB Proceedings, Publication 149. Pp 31-36

Rudnai P., Banhidi L., Sarkany E. Bacskai Z. J., Kertesz M., Martini F., Selmeczy K. (1994) Importance of human and housing factors in the prevalence of SBS symptoms among inhabitants of a residental area in Budapest.

Healthy Building'94 Proceedings of the 3rd International Conference Bp. Vol.1. pp 487-492.

Banhidi L., Kajtar L. (1996) Effect of the External, Air Pollution on Indoor Air Quality and Selecting Mechanical Ventilation System.
Proceedings of Indoor Air'96. Vol 2. Pp 211-216. Nagoya Japan,

Redosi I., Kajtar L., Banhidi L. (1998) Thermal a Comfort in Climatez Office Building in Winter.

Design Construction and Operation of Healthy Buildings, ASHRAE, Atlanta, pp 179-183.

Table 1.

Compulsory subjects

Semester	Subject	Credit	"Contact classes" lecture + seminar + lab.	Home classes	v/f
7.	Heating	3	2+1+0	2	v
	Air conditioning	3	2+0+0	3	v
	Comfort theory	3	3+0+0	2	v
8.	Ventilation	3	2+1+0	2	v
	Heating	3	2+1+0	2	v
9.	Ventilation	3	2+1+0	2	v
	Planing	3	0+1+0	4	f

Optional subjects, but it is compulsory to choose one subject from each group

Semester	Subject	Credit	"Contact classes" lecture + seminar + lab.	Home classes	v/f
7.	Building physsic	2	2+0+0	1	
	Water supply canalization I.	2,5	1,5+0,5+0	2	f
	Building Serv. energetics	2	2+0+0		v
8.	Gas supply	3	1,5+0,5+0	3	f
	Refrigeration	3	2+1+0	2	
	Air cond.systems	3	2+0+0	3	v
9.	Environment protect.	2,5	1,5+0,5+0	2	v
	Building serv.measuring	3	0+0+2	3	f
	Control of build.serv.systems	3	2+1+0	2	v
	District heating	3	3+0,5+0,5	2	f
	Industrial vent.	2,5	1,5+0,5+0	2	v

Optional subjects for building services engineering students to choose for their optional classes

Semester	Subject	Credit	"Contact classes" lecture + seminar + lab.	Home classes	v/f
7.	Mass transfer	2,5	2+0+0	2	
	Applied fluid.mech.	2	2+0+0	1	v
8.	Build.elct.systems	4	3+0+0	3	
	Illumination	3	2+0+0	3	
	Build.serv. solar syst.	2,5	1+0,5+0,5	2	f
9.	Water supply, canalisation	2	2+0+0	1	f
	Energetics simulation of build.	5	3+1+0	3	f
	Simulation of comfort	2,5	1+1+0	2	f

POSTGRADUATE TRAINING ON THE HEALTH SIGNIFICANCE OF INDOOR AIR QUALITY IN HUNGARY

P. RUDNAI

National Institute of Environmental Health, J. Fodor National Public Health Centre
Gyáli ut 2-6, Budapest, H-1097, Hungary

1. Abstract

Health Significance of Indoor Air Quality has been in the curriculum of the postgraduate training courses for doctors to be specialized in public health and epidemiology for almost two decades. In the frame of the postgraduate course on "Ecological architecture" at the Technical University of Budapest it has been taught for two years. Special stress is laid upon the issues that represent current problems of high importance in the everyday public health practice related to the indoor environment. The Hungarian National Environmental Health Action Plan (NEHAP) has also been paying special attention to education relating to the indoor environment.

2. Introduction

It is important that the health significance of a good indoor air quality and the means of achieving it should be well understood by both the architects and civil engineers on the technical side and by the physicians and public health experts on the medical side. This paper presents an overview of the teaching activity on indoor air quality in the postgraduate training courses in Hungary.

3. Post Graduate Training Courses

Health Significance of Indoor Air Quality has been in the curriculum of the postgraduate training courses for doctors to be specialized in public health and epidemiology at the Haynal Imre University of Health Sciences, Budapest for almost two decades. Earlier it covered only the topic of indoor climate and comfort.

In the early '80-ies, however, the importance of indoor air pollution became obvious: more and more artificial materials had been used for building and furnishing both homes and public buildings, which emitted hundreds of chemicals in various amounts; infiltration through the walls of the prefabricated concrete panel buildings was almost zero, especially after windows and doors had been sealed tighter for energy conservation purposes; and, in order to meet the circumstances of the plants producing prefabricated

N. Boschi (ed.), Education and Training in Indoor Air Sciences, 81–83.
© 1999 *Kluwer Academic Publishers. Printed in the Netherlands.*

concrete elements, the minimum height of the rooms had been decreased to 2.50 m. As a result of all these changes, the pollutants emitted into the indoor air cumulated and reached soon a critical concentration with possible adverse effects on the health of the inhabitants. People, indeed, started to complain for experiencing certain non-specific symptoms, later known as the sick building syndrome. The results of our domestic [1] and international [2,3] research collaborations helped to fill in the lectures with materials relevant to the Hungarian situation.

In the frame of the postgraduate course on "Ecological architecture" at the Technical University of Budapest the health aspects of indoor air quality has been taught for two years. Engineers, medical doctors, biologists and other scientists are participants of this latter course.

In both courses the topic of indoor air is embedded in a broader field of *health significance of indoor environment* (besides indoor air, noise, vibration, electromagnetic field, insolation, lighting, harmony between structure and function, overcrowding etc. are covered).

The chapter on *health aspects of indoor air quality* consists of the following parts:
1. Health significance of indoor air quality
   - Temperature and humidity (physiological background of thermoregulation, sources of microclimatic problems, health and comfort problems, hygienic requirements)
   - Ion content (physiology, health effects, hygienic requirements)
   - Chemical pollutants in indoor air (sources, effects and control).
   - Microbial pollution (health effects, hygienic requirements)
   - Sick Building Syndrome (causes, medical symptoms, confounding factors)
2. Emissions from building materials and their control
3. Environmental Tobacco Smoke
4. Indoor allergens

These topics are covered by three lectures at the Haynal Imre University of Health Sciences and by 12 – 16 hours of lectures at the Technical University. In the latter case we need to present the physiological basis more detailed and the health effects have to be better explained as well.

Under the subheading of "Chemical pollutants in the indoor air" special stress is laid upon those pollutants and their sources that cause a lot of problems in the everyday public health practice in Hungary. One of them is the gas heaters which are not connected to a chimney but the exhaust gas is lead out to the open air just below the window seal and this way most of the pollutants are working their way back into the rooms. Measurements showed that the concentration of nitrogen dioxide (NO_2) around a high-rise building equipped with such gas heaters is well over the hygienic limit value. According to our earlier measurements [4], the NO_2 level is about 30-40 $\mu g/m^3$ higher in flats with such gas heaters than with gas cookers only. Children living in flats with such gas heaters have a 50 % higher chance to get ill with acute laryngitis than those who live in flats with central heating. It will take some time, however, until, as a result of education and training, all engineers (and inhabitants!) will accept that gas heaters should be connected to chimneys.

The importance of education and training in the provision of a healthy indoor environment was accepted and stressed by the Hungarian National Environmental Health Action Plan, too, in the following points:
- Engineers, family doctors and the population should be informed on the health aspects of housing
- Education of proper behaviour in the indoor environment should be promoted

References

1. Rudnai P, Sárkány E, Bánhidi L et al, 1994, Importance of Human and Housing Factors in Prevalence of SBS Symptoms Among Inhabitants of a Residential Area in Budapest, *Proceedings of Healthy Buildings '94, Budapest*, Vol 1, pp.487-492
2. Thielebeule U, Farkas I, Hülsse Ch, Rudnai P, 1989, Indoor Air Pollution of Formaldehyde in New and Old Buildings and Health of Children, *Zbl. für Hyg. und Umweltmed*, 189/1
3. Rudnai P, Farkas I, Bácskai J, Hülsse C, Sárkány E, Dávid A, Thielebeule U, 1990, Epidemiological Study on Indoor Formaldehyde Exposure of Children, *Proceedings of Indoor Ar '90, Toronto, Canada*, Vol. 1, pp. 459-464
4. Rudnai P, Farkas I, Bácskai J, Sárkány E, Somogyi J, 1993, Indoor NO_2 Levels in Homes with Different Sources of Air Pollution (Traffic, Gas-use, Smoking), *Proceedings of Indoor Air '93, Helsinki, Finland*, Vol 3, pp. 195-198

INDOOR AIR QUALITY EDUCATION AT THE SLOVAK UNIVERSITY OF TECHNOLOGY IN BRATISLAVA SLOVAK REPUBLIC

DUSAN PETRAS
Slovak Technical University
Slovak Republic

Introduction

The Slovak University of Technology in Bratislava was founded in 1937. It consists of the following 5 faculties:
1 Civil Engineering Faculty
2 Mechanical Engineering Faculty
3 Faculty of Architecture
4 Faculty of Chemistry
5 Faculty of Electrical Engineering

The subject of Indoor Air Quality is taught at the Civil Engineering Faculty and the Faculty of Mechanical Engineering.

1. Teaching at the Civil Engineering Faculty

1.1 STRUCTURE OF THE FACULTY

The Civil Engineering Faculty is the second largest faculty at the Slovak University. With its 370 teachers and researchers, 22 vocational and theoretical departments and laboratories, computer centre and library, it represents a significant educational and scientific research centre in Slovakia. The Faculty enjoys partial legal and economic independence within the University. The freedom of teaching and research is constitutionaly guaranteed. The democratically elected Academic Senat (Faculty Parliament) composed of deputies representing teachers, researchers, employes, postgraduate and undergraduate students, together with the independent Scientific Council of the Faculty, provide the necessary control and feedback mechanism for the management of the Faculty.

At the present time, there are approximately 2 900 students at the Faculty of Civil Engineering who attend courses in some of the four main fields of study:
- architecture and building construction,
- structural engineering and transportation engineering,
- water resources management and engineering,
- geodesy and cartography.

N. Boschi (ed.), Education and Training in Indoor Air Sciences, 85–92.
© 1999 *Kluwer Academic Publishers. Printed in the Netherlands.*

Graduates of the broad field of architecture and building construction are trained to construct building sand engineering works that respect ecological, sociological, and engineering principles. They are able to solve tasks in architectural design and structural construction, the technical equipment of buildings, building materials engineering and the functional implementation of buildings as well as financial and construction management, quality management and inspection.

1.2 BUILDING SERVICES SPECIALIZATION

Building services study in under the quarantee of the Department of Building Services

I.   STAFF

      Professors - 2
      Associate Professors - 3
      Lecturers - 12
      Research Fellows - 2
      Postgraduate Students - 4
      Technical Staff - 2

II.   EQUIPMENT

II.1   Teaching and Research Laboratories
      Laboratory in Trnavka

II.2   Special Measuring Instruments and Computers
      Bruel & Kjer Thermal Comfort Analyzer
      Anemometer
      Izomet 104
      Globemeter
      Measuring Unit

      Computer room:
      PC computers: DTK 486, Escom 368, Eurocomp-Pentium
      Plotter
      Printer

III.   TEACHING

III.1   Graduate Study

Subject	semester	hours per week	
		lectures	seminars
Technical Equipment of Buildings I	4	2	2
Technical Equipment of Buildings II	5	2	1
Technical Equipment of Buildings III	6	2	1
Internal Water and Gas Pipelines	6	3	2
Internal Drainage	6	2	2
TEB Machine Equipment	6	1	1
Ventilation	7	2	2
Heating I	7	2	2
Heating II	8	2	2
Design IV	7	0	4
Design V	8	0	4
Design VI	9	0	3
Design VII	10	0	6
Climatization	8	2	2
Computer Design Systems	8	0	3
Measuring and Regulation I	8	2	2
Measuring and Regulation II	9	2	2
Energy Supply of Buildings	9	2	2
Excursion	9	2	2
Building Control Systems	10	3	2
Special Seminar	10	0	2
Energy Saving in Buildings	10	2	2
Technical Equipment of Buildings	3	1	1
Heating Systems	9	2	2
Industrial Installations	9	2	2
Renewable Energy in TEB	9	2	2
Ventilation & Climatization Systems I	9	2	2
Ventilation & Climatization Systems II	9	2	2
Electrical Installations III	9	2	2
Research Methods in Sanitation	10	2	4
Research Methods in Heating	10	2	4
Research Methods in Ventilation	10	2	4

IV.   RESEARCH TARGETS

- possible water savings via technical devices
- lowering energy consumption in buildings
- indoor climate of buildings
- modernization of heating systems
- air and moisture energy transfer

V.   RESEARCH PROJECTS

1. Energy and Environmental Access in Choice of Progressive Heating Systems in Buildings (3 years, Assoc.Prof. Petráš, D.)
2. Environmental and Energy Evaluation of Sanitary and Airconditioning Systems for Dwelling Renovations (3 years, Prof. Valášek)
3. In cooperation with SAV-USTARCH: Energy, Air, and Moisture Transfer in Heterogeneous
Perimeter Construction (3 years, Dr. Matiašovský)

2. Teaching at the Mechanical Engineering Faculty

2.1 STRUCTURE OF THE FACULTY

In 1996, our faculty graduated 96 BSc students and 173 MSc students, the PhD degree was aworded to 17 young research workers. In the academic year 1996/97, 2 136 students were enrolled in all forms of study , 284 in BSc, 1 852 in MSc, and 126 PhD studies respectively.

Nowadays the faculty offers four fields of undergraduate study:
- Management of Machine Production
- Automation and Engineering Informatics
- Motor Vehicle Engineering
- Environmental Engineering

and eleven fields of graduate study:
- Thermal Power Engineering and Environmental Technology
- Hydraulic and Pneumatic Machines and Equipment
- Production Systems with Industrial Robots and Manipulators
- Machines and Equipment for Engineering Production
- Machines and Equipment for Chemical, Food and Consumer Industry
- Machines and Equipment for Civil Ehgineering, Land Reclamation and Agriculture
- Transport Technology
- Instrumentation, Informatics and Automation Technologies
- Applied Mechanics
- Engineering Technologies and Materials
- Management of Machinery Production

The faculty also offers the postgraduate studies (PhD studies for Slovak and English speaking students, linked with the research programme of particular departments in the following study fields:
- Transport Technology
- Power Engineering, Machines and Equipment
- Machine Parts and Mechanisms
- Engineering Mechnics
- Measurement and Control of Machines and Processes

- Production Technology
- Process Engineering
- Engineering Technologies

Research activities cover all the areas of science and engineering mentioned above. In the past year the research programme included 6 international projects, 21 projects financed by domestic grant agencies and more than 100 scientific, research and development projects of national or local significance, suporting mechanical engineering technology and production.

2.2 HEAT ENGINEERING SPECIALIZATION

I. STAFF
Associate professors- 6
Assistant professors – 3
Research workers – 4
PhD students – 7
Technical staff – 3

II. EQUIPMENT

II.1 Teaching and Research Laboratories

- Laboratories of the thermodynamics (thermodynamics and heat transfer)
- Laboratory of the two-phase flow problems
- Laboratory of the aircondition problems
- Refrigerating engineering laboratory
- Low temperature laboratory

II.2 Special Measuring Instruments and Computers:

- PC software FLUENT V 4.2 (chemical and process eng.); component design, combustion design and engineering, aerodynamic design, electronics cooling, power generation, heat transfer operation
- Inteligent temperature and pressure sensor SMART Honeywell
- Jet mills for desintegration of solid materials
- Low temperature calorimeter for solid materials thermal conductivity and heat capacity
- heat pumps
- SULZER dew point meter
- TESTO 452 (Testoterm) temperature, pressure, flow and humid meter
- JUMO C 200 registering apparatus (6 points)

III. TEACHING

III.1 Graduate Study

Name of subject	semester	hours per week	
		lectures	seminars
Thermodynamics	5	3	2
Heat Transfer	6	2	1
Heat and Mass Transfer	7	3	2
Environmetal Technique	7	3	2
Heat Pumps and Low Potential Energy	8	2	2
Compressor and Refrigerating Engineering Technique	7	3	1
Industry Ventilation	7	2	1
Basic Theory of Ventilation and Air Conditioning	8	3	2
Experimental Methods	7	3	2
Aircondition	9	2	2
Airprotection Technique 1	9	2	1
Airprotection Technique 1	10	2	1
Design, Working and Production of Environmental Devices	9	2	1
Refrigerating Engineering	8	2	1
Tests of Machines	7	2	2
Aircondition in Industry	8	2	2

III.2 Postgraduate Study

Special Thermodynamics Problems	4	36	
Theory of Thermodynamics Systems	3	36	
Thermal Systems and Equipments- Experimental and Diagnostic Methods	4	8	

IV. RESEARCH TARGETS

Aircondition theory
- creating a new calculating methods and objective standards for energetic optimal HVAC systems
- nonstationary heat transfer throug walls
- computer aided air – conditioning systems design

Thermal and refrigerating systems
- effect of the condenser performance
- energy efficiency in refrigeration
- primary heat sources and cogeneration units problems

Environmental problems connected with refrigeration systems
- absorbtion refrigeration systems
- CFC free refregeration - a new thermodynamic equations of state
- alternative refrigerants

Multiphase flow theory
- resistance coefficient determination connetcted with concentration solids particles in flow

Low temperature thermodynamics
- thermophysical properties at low temperatures
- mathematical model of multilayers insulation

V. RESEARCH PROJECTS

Faculty projects
- Research of absorbtion cooling systems and heat pumps.
- Leader: K.Mecárik
- A complex utilization of fuel primary energy in combined energetic equipment. Leader: L. Pastor
- Energetical co-generation system with heat pumps.
- Leader: V. Havelský

VEGA grants
- Ecological and economical utilization of heat for obtaining different energy forms including cool. Leader: L.Pastor
- Evaluation criterions for design of energetically optimal systems for interior clima.
- Leader: K.Ferstl
- Utilization of solar energy for heating, cooling and air-conditioning.
- Leader: V.Havelský
- Research and development of non-azeotropic mixtures and their application in thernmal and cooling cycles. Leader: K.Mecárik

Contractor s activity
- Software program Racio III.
- Leader: Š. Antal
- Energetical system for production of electric and heat energy by combined co-generation units with heat pumps.
- Leader: V.Havelský
- Adsorption combined air-conditioned unit powered by gas.
- Leader: M.Horák
- The utilization of the earth gas in airconditioning technique.
- Leader: M.Masaryk

4. Conclusions

Based on explanation of the education at the Civil Engineering and Mechanical Engineering Faculties, it is clear that the subjects Indoor Environment = Indoor Air Quality +Climate are tought indirectly as the part of the technical disciplines e.g.: heating, ventilation, airconditioning, cooling....

ENVIRONMENTAL ENGINEERING EDUCATIONAL PROCESS IN THE FIELD OF INDOOR AIR SCIENCES

INGRID SENITKOVA
Head of Civil Environmental Institute
Technical University Kosice, Slovakia

1. Introduction

As education in the field of indoor air sciences in Slovakia, is a part of environmental education of Civil Engineering Faculty, Technical University Kosice, let me propose some basic considerations that may suggest a new view for understanding of relationships between civil and indoor environmental engineering. We believe that our moral responsibility to coming generations for civil environmental education especially in indoor air problems will follow directly from these thoughts.

The indoor environmental engineering is to be so urgent for civil engineering education as two important aspects do exist. The problem of indoor is mostly the problem of interaction between engineering construction and surroundings or indoor-outdoor building construction interaction. Both of them have to be solve as technical problem with a lot of engineering and environmental feelings. By the way we spend our time mostly indoors, so the indoor environmental problems haven' t been forgotten. So the answer on the question why it is necessary to implant the indoor environmental education into civil engineering educational process is based on the fact that indoor air problems are mostly connected with civil engineering constructions generally. As indoor is artificially created internal environment connected to building construction it is clear that education in this subject is to be course of civil engineering study. These facts explain us the interaction of civil and indoor environmental engineering.

2. Environmental education

What are the reasons for the low environmental awareness and participation of the population relative to the also unsatisfactory level of environmental awareness. It was, and still is, generally accepted that the main reason for the ignorance is lack of information about the real state of the environment. As well, materials on environmental protection are much more available today. Maybe the reasons are not much lack of information, but rather lack of alternatives and independent thinking, lack of motivation to act, distrust in the possibility to influence environmental policy, and emphasis on material consumption as the expression of social status.

N. Boschi (ed.), Education and Training in Indoor Air Sciences, 93–96.

The environmental education has to have the close connection to environmental science. To understand the term, take each word separately. The word environmental refers broadly to everything around us. Science refers to a body of knowledge about the world and all its parts. It is also a method for finding new information. Science seeks exactness through measurement, insight through close observation and foresight through its theories. Environmental science came into existence as a recognized discipline and is aimed as helping us control our own actions in the natural world to avoid irreparable damage. In this sense, environmental science means learning to master ourselves.

We welcome to a new kind of science because it helps us to solve the highly complex problems of many scientific fields. Spanning suck a wide range of knowledge, environmental science offers an integrated view of the world and our part it Environmental science take on the colossal task of understanding complex issues and it is the study of the environment its living and nonliving components, and the interactions of these components.

3. Civil environmental engineering study course

The Civil environmental engineering study course at the Faculty of Civil Engineering might be made up of so-called basic study which usually lasts five years. This is a basic for the next part and emphasized fundamental natural sciences. The next part of study is concentrated on basic subject profiling civil engineers with respect of special environmental problems. The study of special subjects with environmental orientation starts in the third study year, when the special subjects focused on typical environmental fields concerning to civil engineering are taught.

One of the main subjects studied at the environmental study course is subject Civil Engineering Ecology. Its goal is to explain to students the basic global environmental problems. The next one is Environmental Engineering. Its goal is to show the possible methods of environmental problems solving from technical point of view. The other following subjects as Protection of water, Air meteorology, air quality, Erosion and Solid waste help to gain general necessary environmental knowledge. Alternative energy sources, indoor air quality should be helpful for making solution in indoor environmental problems. Student's knowledge are deepened and broadened by study of Theory of internal microclimate, HVAC and Water engineering also. The special attention during the study is paid to teaching of monitoring and management of environmental processes, as well as to law and legislation rules. The study of environmental engineering course is focused on the education of a civil engineer, who is able on the basis of gained knowledge, to work on progressive projects and realization of civil engineering structures. With the help of deepened knowledge in the field of ecology and environmental protection he is able to find efficient and suitable solutions of complicated environmental design problems. His education in this field allows him to reduce or even remove the negative influence on environment generally.

4. Indoor environment study

Special part of Civil environmental engineering study course on Civil Engineering Faculty Technical University Kosice is Indoor environment study program where the problems of indoor air are included. Indoor environment is analyzed from physical, chemical and biological point of view. Presented building air quality study course is based on what is known and generally accepted at this time in the relevant fields of building sciences and in the field of indoor air quality. The purpose of the study is the civil engineers' ability to design good indoor environment quality, which means and includes mainly maintenance of acceptable indoor environment from building design, implementation and operation process point of view.

Indoor environment graduate study contents parts concerning to intelligent buildings with automatic service control. All study efforts are concentrated to obtain the skills required to design healthy buildings as well as to design low-energy buildings. All these design strategies are completed with skills required to creates good IAQ because good indoor environment has to have good indoor air quality.

5. Indoor air study

The content of the lecture on Indoor Air sciences are focused on the topics:

- Indoor Air Quality (building physics, building chemistry and building ecology).

- Factors Affecting IAQ (source of indoor air contaminants, HVAC system design and operation, pollutant pathways and driving forces and building occupants.

- Building Measurement, Diagnosis (thermal comfort, moisture, mould and mildew, radon, asbestos, man - made mineral fibres, dust, volatile organic compounds, nitrogen oxides, carbon monoxides, voice etc.).

- Preventing IAQ Problems (indoor environmental law, developing an IAQ management, tools required to create good IAQ).

- Resolving IAQ Problems (using occupant data, the HVAC - system data, pollutant pathway data, pollutant source data, collecting additional information, design or re-design of the buildings).

Besides the indoor environment graduate study Civil Environmental Institute on Civil Engineering Faculty of Technical University Kosice offers postgraduate level programs in the field of environmental engineering.

These programs cover all professional areas concerning to civil environmental engineering especially to air protection, water treatment, waste management, building ecology, indoor environment, environmental risk assessment, environmental management, etc.

Indoor air sciences are trained in the postgraduate programs in the field of indoor environmental physics and chemistry (thermal comfort, moisture, radon, VOC_s, nitrogen oxides, filnous dust). Nowadays at the Civil Environmental Institute these postgraduate study programs in the Indoor air sciences are going:

- Indoor Radon Levels and Ventilation Rate
- Indoor VOC_s Concentrations and Exposures
- Nitrogen Oxides Impacts on Indoor Air Quality
- Indoor Fibrous Dust Concentrations and Exposure.

6. Conclusion

Indoor air sciences educational activities are particulars implement to Civil Engineering study in Slovakia by various ways. The level of professional knowledge in this field depends on the type of study course. The indoor environment lectures are also part of other civil engineering courses study as building construction or technical services for buildings. Today no indoor environment science does exist. There is no possibility to get a degree in such a discipline. You could still make a career with in civil environmental engineering discipline dealing with technical, hygienic, chemical and other issues. There is needed much more interdisciplinary efforts. What is really needed is a new parodign in indoor environmental sciences especially in indoor air and an internationally harmonized study programs in this field.

INDOOR AIR EDUCATION IN THE SLOVAK REPUBLIC.

Education at the medical faculties

KATARINA SLOTOVA M.D.
State Health Institute
Cesta k nemocnici 1
975 56 Banska Bystrica
Slovak Republic

1. Historical Development

A comprehensive system of medical teaching with an emphasis on the preventive aspects of medicine was developed at the Medical faculty of hygiene of Charles University in Prague in 1953.

The curriculum included classical medical topics, as well as special topics about preventive medicine including environmental factors and their impact on public health but without special attention to indoor air education.

There was only one Faculty for the whole Czechoslovak Republic and their graduates – medical doctors created a good base of experts in the field of environment and health.
Since November 1989, many organizational changes have been aimed at the introduction of a new study programme. The name of the Faculty has been changed from the Medical Faculty of the Hygiene to the Third Medical faculty of Charles University in Prague. The curriculum was amended so that it would correspond to the general orientation of the Faculty of medicine using the previously developed structure from the branches of preventive medicine.

2. Present Situation

After the separation of the Czechoslovak Republic, there are only three medical faculties and one nursing and social work faculty in the Slovak Republic:
- Comenius University Bratislava-Medical Faculty
- Safarik University Kosice-Medical Faculty
- Medical Faculty - Martin

97

N. Boschi (ed.), Education and Training in Indoor Air Sciences, 97–99.
© 1999 *Kluwer Academic Publishers. Printed in the Netherlands.*

- University Trnava-School of nursing and social work students at each medical faculty are encouraged to develop their knowledge in various elective subjects including Environmental health.

These medical institutes have played a leading role in the education and training in the field of Indoor Air and its health effects.

The non-industrial indoor air research program on the impacts on public health has been established in Slovakia mainly during the last ten years. But the educational and training program has not been developed simultaneously at the appropriate level and it is still inadequate.

2.1. OVERVIEW OF A SPECIFIC SUBJECTS RELATED TO GRADUATE EDUCATION.

In the subject of Hygiene are among other involved topics (for students in 4-year study, 3 lectures) about:
- Health and environment
- Environmental impact of the chemical, physical, biological, and psychosocial factors on health
- Outdoor and Indoor air pollution and health effects
- Ionizing radiation and health hazards
- The impact of noise on health
- Urbanization and health principles of housing
- Hospital hygiene, outpatient departments, and inpatient units
- Health protection against radiological, chemical, and biological agents

Contents of the practical training- four lectures:
Radon in homes and its influence on human health
Hygiene of health facilities:
- Outpatient departments
- Hospitals, inpatient units, and wards
- Environmental microclimate, the effect of acoustics on health , and microbiological contamination
- Inpatient unit: the prevention of nosocomial infections and the hygienic investigation of the inpatient unit

2.1.1. *Other activities in the field of education related to Indoor Air.*

Since 1995, the U.S. Agency for International Development (US AID) has been working in eastern Europe to help address the health consequences of environmental pollution. Through the Environmental Health Project (EHP), a prior project has been conducted which focused on strengthening the research capacity and risk communication methods used by state health agencies and municipal governments to address pollution problems affecting human health.The present project extends environmental health activities in Slovakia into the area of curriculum development.

In the project are involved:

Technical University-Faculty of Environmental Sciences in Zvolen

Matej Bel University-Faculty of Economy and Faculty of Natural Sciences in Banska Bystrica

The elaborated course curriculum for the education at the mentioned faculties includes topics: Living Environment (2-3 lectures)

A: Educational Objectives are:

Explain the term, living environment - urban and rural

Understand how the quality of the living environment can affect health

Describe how public officials can influence the health of the living environment.

B: Key content:

Introduce environmental factors that affect health in living environment

Differences in health risk due to the environment in urban and rural areas

Sources of risk and methods to control them.

Postgraduate educational Programme: Postgraduate educational programs have not been done very successfully in the field of health and Indoor Air sciences for physicians, epidemiologists, and exposure experts. Postgraduate education is more developed only in the field of ionizing radiation, hospital environment, and prevention of the nosocomial infections. Physicians can develop other Indoor Air information only on their own initiative or in the framework of international projects, about Environment and health. Special attention is usually paid to creating investigative methods, building materials, radon mitigation, maintenance of ventilation systems, moisture in buildings, mold growth, allergies, and infectious diseases.

3. Suggestion for the Future

Many of the actions to improve indoor air quality are dependent on those who are designing, using, and maintaining buildings. Therefore, we should address the education and training of profesionals and other decision makers, including health professionals.

We should also include a program for the dissemination of information to the public in order to satisfy people's "right to know" and stimulate changes in behaviour.

For the future, it is necessary to create more extensive graduate and postgraduate programs in the field of indoor air and its impact on health, to educate new professional profiles from the health sector, and make professionals from the educational sector aware of the importance of Indoor Air Quality and their responsibility in this field.

PART V. RESEARCH AND PROBLEM BASED EDUCATION

TRAINING SPECIALISTS IN OCCUPATIONAL MEDICINE: THE BENEFITS OF RESEARCH EXPERIENCE IN THE FIELD OF INDOOR ENVIRONMENT

G. MUZI

Occupational Medicine and Occupational and Environmental Toxicology - University of Perugia, Perugia (Italy)

1. Introduction

In Italy there is no formal course on "Indoor air sciences" in the undergraduate curriculum of the Faculty of Medicine, which lasts for 6 years. A few aspects of "Indoor environment and health" are taught in the course on Public Health and most of the basic information is provided during the Course on Occupational Medicine.

Students learn the main features of the most frequent illnesses caused by indoor pollutants and some diagnostic procedures. Preventive measures and basic concepts such as building ecology, ventilation, source control and exposure are barely mentioned. Moreover, little attention is paid to long term effects of some indoor pollutants such as radon, benzene, formaldehyde, fibres, etc.

Given this background, over the past decade we have been trying to promote a general interest in Indoor Air Quality, mainly in the working environment, in the post-graduate school in Occupational Medicine. As most people in developed countries now work in service industries such as offices, hospitals, schools, etc., the specialist in occupational medicine will very likely be responsible for the health and comfort of these groups of workers operating indoors. Furthermore, at present in Europe one of the duties of the specialist in Occupational Medicine is to inform the workers in his care on how to cope with environmental risk factors, including indoor air pollutants, and to prevent their adverse health effects.

This paper outlines when and for how long research activity should be performed by the trainee specialist in Occupational Medicine, what type of study and which are the main areas of research which are most beneficial and why the trainee specialist will find research experience in the field of Indoor Environment important for his/her future career.

N. Boschi (ed.), Education and Training in Indoor Air Sciences, 103–107.

2. Research experience: when and for how long

In Italy the postgraduate specialist Course in Occupational Medicine lasts for four years, and includes teaching on ergonomy, physiology, toxicology, prevention, epidemiology, clinical training in occupation-related diseases, forensic medicine, emergency, audiology, dermatology, etc. Furthermore, trainee specialists must take part in at least three "controlled clinical studies", but no other specification on research activity during training is provided by law.

The best time to perform research activity is during the post graduate training course. In fact, young doctors who have just graduated have little practical experience and if they were immediately involved in a research project they might encounter difficulties when faced with clinical practice later.

On the other hand, carrying out a research project after finishing specialist training means a substantial loss of income [1]. In contrast, during his/her specialist training the doctor is already skilled enough in the practice of Occupational Medicine and is strongly motivated to return to practical activity whether clinical or preventive when it ends.

The ideal period of research is one year, and since the type of study may involve a follow up, the 12 months research may be divided into sessions of 3 or 4 months per year, starting in the second year.

3. Research in the field of Indoor Environment

Choosing a research field is of primary importance. Moreover, deciding on a research question is often difficult for the specialist in training and the support of an experienced researcher is fundamental. In practical terms the specialist in training usually takes part in an on-going research project in the institution he is attending for his post-graduate diploma. Some examples of areas of research that are known to be of importance in Occupational Medicine are musculo-skeletal disorders of the back and upper limbs, asthma, accidents, skin disorders, vibration induced diseases, suicide and depression, and finally hearing loss [2]. Other priorities are defining occupational health screening procedures, evaluating the impact of industrial activity on community health, and investigating the neuropsychological aspect of work exposure with an emphasis on diagnostic tests [3]. Indoor Environment has recently been defined as a priority research area in a statement of the CDC/NIOSH [4]. Therefore, research evaluating disturbancies or diseases, mainly due to poor Indoor Air Quality, in workers in confined non-industrial environments, is of special interest for the trainee. In fact, many questions in this field still remain to be answered.

- We still need to establish the aetiology of disturbances or diseases such as for example the Sick Building Syndrome and the Multiple Chemical Sensitivity as well as the importance of psychosocial and non work-related factors
- The short and long term effects of mutagenic and cancerogenic factors, such as the Environmental Tobacco Smoke and Radon, are not completely known and many

cause-effect relationships, in particular with regard to mixture of pollutants at low doses, have to be investigated.
- Strategies for diagnosing and preventing some building-related illnesses need to be better defined.
- The impact on health of different climates, ways of working, building techniques, conditioning systems and furnishings needs to be assessed; the equipment and procedures for environmental evaluation must be updated when investigating work-places with indoor air problems.

4. Type of research and methodology

Proper definitions of the objectives of a study, the type of research and/or the methodology are also very important as they prevent the project from becoming a dreary unmotivated activity [5]. The trainee specialist should have ample opportunity to discuss objectives, feasibility and methodology with experienced tutors.

As basic research usually uses tools and methods which are never applied in daily practice, epidemiological studies are most useful for the future specialist whether they are observational (cross-sectional, case-control or cohort studies) or interventional (controlled clinical trials). In *cross-sectional* surveys the potential causes and the outcome are measured at the same point in time[6][7]. They are usually applied in the field of Indoor Environment. However, as they are relatively rapid and inexpensive to perform, they can be easily applied in any kind of future working situation.

Whenever possible the specialist in training should be involved in *cohort studies*. They have the major advantage that the potential causes under study precede the outcome. Therefore, the cause effect relationship is evaluated accurately. Curing for cohorts of workers will probably be the full time occupation of most future specialists, and learning to evaluate biological or pathological end-points appropriately and to correlate them correctly with the working exposure is obviously of great benefit.

In *clinical trials*, which are needed in the Indoor field, an investigator randomises individuals to study and control groups and then intervenes in some way. An experience with this kind of study may be helpful for the future specialist when he has, for instance, to evaluate the outcome of an environmental modification [6].

5. Aims of research training

The covert aims of any research training are to ensure the trainee acquires [1]:
- a sound scientific method;
- a critical awareness of the tools of the profession, together with an ability to understand new developments in his field and finally,
- adequate analytical and communication skills.

The practice of Occupational Medicine implies adequate knowledge of scientific methodology. In his daily job the specialist reaches far beyond the application of general principles to interpret signs and symptoms, to formulate a diagnosis and to prescribe treatment. He is often called upon to hypothesize, establish and prove a cause-effect relationship between potential or actual pathological events and risk or hazard factors. Consequently, the specialist in Occupational Medicine requires training in how to formulate hypotheses, and how to test them using the three criteria of contributory cause i.e. in the first place the characteristics referred to as the "cause" are associated with the disease (effect), and occur more often than expected by chance alone in the same individual as the disease; secondly, that the cause precedes the effect, and finally that altering only the cause alters the probability of the effect (disease) [6].

Experience in research is essential for acquiring the tools of the profession, i.e. the ideas and techniques which will be properly used throughout the specialists' working life. This cannot be limited to either the single occasion of the post-graduate training and study period or to the transmission of proven concepts. Technological progress constantly provides new instruments which need to be chosen critically and selected with care [1]. If one considers only the diagnosis of building-related illness such as bronchial asthma and alveolitis or cancer and neuropathies, one can see how the tools of the profession have changed greatly over the past ten years. Examples include studying bronchial reactivity when diagnosing professional asthma, using high resolution thoracic CT scans to visualise early interstitial lung damage, and using batteries of computerised tests to assess slight alterations in neuro-behavioural patterns.

Evidence is even greater in health surveillance and biological monitoring to prevent occupationally related illness [8][9].

Consequently, specialists need to be educated towards an independent, unbiased understanding of scientific progress. This understanding cannot be limited to a critical analysis of any and all scientific reports, but must be based on a sound scientific background and a practical knowledge of research skills which will ensure the specialist understands and recognises scientific arguments which can successfully be applied in practice.

Having good analytical and communication skills, is essential for the specialist in Occupational Medicine because often he/she has to work in team with other specialists, cope with the management of a company, with the health and safety officer, with workers representatives, the workers themselves, government health inspectors, and liaise with general practitioners and physicians in local hospitals [3].

Again one practical example is the recent involvement of specialists in Occupational Medicine in the medical prevention of building-related illness and complaints. In this case, as a multidisciplinary multi-step approach is the most appropriate, the specialist in Occupational Medicine must be ready to participate in and perhaps co-ordinate the work of a group of experts including specialists in industrial hygiene, epidemiology, plant engineers, chemists and microbiologists.

6. Conclusions

How does being involved in a research project in the field of Indoor Environment as a trainee specialist help develop the qualities required?

In the first place this research develops analytical skills to high degree; the trainee learns to support his argument with clear evidence, to develop and organize a line of thought and last but not least, to read between the lines of a scientific report. The achievement of scientific leadership is much easier after a solid research experience has provided adequate analytical, organizational and communication skills, besides the practice of working in team for the same target.

In conclusion, research experience during the post graduate training course is extremely important because it educates a class of specialists who are capable of a critical understanding of medical progress, who can evaluate its practical applications, and finally who have acquired and who will continue to develop criteria which are indispensable for their professional career. Research in the field of Indoor Environment may provide a unique opportunity to be involved in a multidisciplinary project and to be at the cutting edge of Occupational Medicine.

7. References

1. Editorial (1993) Does research make for better doctors? *Lancet* **342**, 1063-1064.
2. Harrington J M. (1994) Research priorities in occupational medicine: a survey of United Kingdom medical opinion by the Delphi technique. *J. Occup. Environ. Med.* **51**, 289-294.
3. Seaton A and Agius R. (1995) Occupational Medicine: the way ahead. *J. Occup. Environ. Med.* **52**, 497-499.
4. de la Hoz, R.E. and Parker, J.E. (1998) Occupational and environmental medicine in the United States. *Int. Arch. Occup. Environ. Health* **71**, 155-161.
5. Riegelman R K and Hirsc R P. (1989) *Studying a study and testing a test*. Little, Brown and Company, Boston.
6. Hennekens C H and Buring J E. (1987) *Epidemiology in medicine*. Little, Brown and Company, Boston,.
7. Muzi, G., Abbritti, G., Accattoli, M.P. and dell'Omo, M. (1998) Prevalence of irritative symptoms in a non-problem air-conditioned office building. *Int. Arch. Occup. Environ. Health* **71**, 372-378.
8. Lauwerys R R and Hoet P. (1993) *Industrial chemical exposure*. Lewis Publishers, Boca Raton.
9. Abbritti, G. and Muzi, G. (1995) *Indoor air quality and health effects in office buildings*. Proceedings of healthy buildings '95, **1**, 185-195.

INDOOR ENVIRONMENTAL QUALITY RESEARCH AND EDUCATION AT HARVARD UNIVERSITY

H.S. Brightman, M.W. First, J.D. Spengler
Harvard School of Public Health
665 Huntington Avenue, Boston, MA 02115
howard@hsph.harvard.edu, fax (617) 432-4122

"The Harvard School is prepared to offer courses of instruction in industrial hygiene and facilities to investigate the problems of industry." (President A. Lawrence Lowell, 1918)

1. Problem Statement

Harvard University was chosen as one example of the development of research and graduate education in the field of indoor environmental quality (IEQ) at a major United States (US) university. Indeed, it was the first institution in the world to establish a course of instruction and research in industrial hygiene. Harvard has a long history of research in IEQ, first through its strong research in industrial hygiene and later through its combined focus on ambient and indoor air pollution. As such, the Harvard School of Public Health (HSPH) provides an educational paradigm that has advanced healthful indoor environments.

2. Objectives

This paper has three goals: 1) to describe the advancement of IEQ at HSPH, including some of the unique achievements and personalities in its history; 2) to provide a snapshot of the current IEQ research at HSPH; and 3) to outline the curricula graduate students and professionals draw upon at the school.

2.1 THE INDUSTRIAL HYGIENE PIONEERS

IEQ research began in 1913 at Harvard University as part of the Harvard-Massachusetts Institute of Technology *School for Health Officers*. This research was performed in response to health concerns about the quickly expanding industrial sector. As a result, the first curriculum included industrial hygiene and sanitation. These courses covered the hazards known to be associated with industry at the time: occupational injuries, industrial poisoning, and the effects of ventilation and dusty trades on tuberculosis and other diseases.

N. Boschi (ed.), Education and Training in Indoor Air Sciences, 109–122.
© 1999 *Kluwer Academic Publishers. Printed in the Netherlands.*

The onset of World War I in 1917 dramatically expanded the scope of research in hazardous work environments. As a result, then Harvard president A. Lawrence Lowell implemented plans to organise a *Division of Industrial Hygiene*. Despite the difficulties presented by the war, ten students were instructed in the division's first year. Dr. Cecil K. Drinker was then placed in charge of the division in November 1918.

An unexpected by-product of this new program was a flood of inquiries from plant managers; those questions that could not be easily answered became the research endeavours of the faculty. One of the first projects was the investigation of poisoning at the New Jersey Zinc Co; manganese was proven to affect the neural ganglia. This research was the first of an extended series of investigations on dust and dust hazards.

Perhaps the most extraordinary personality at the *Division of Industrial Hygiene* in 1919 was Alice Hamilton. Dr. Hamilton was the first woman to be appointed professor at the

university, preceding the entrance of women students by 26 years. At a recent stamp commemoration for "the Great American" series, Dean Harvey Fineberg described her as "the first physician to use the scientific approach to study threats to health in the workplace". Throughout her 69-year career as an industrial toxicologist and social reformer, she worked tirelessly to promote workers' health, dramatically reducing their exposure to such toxins as lead, radium, and mercury. One of her central research themes was the chronic aspects of carbon monoxide. Of this, she commented: "I could not prove it [to be] chronic. You just could not make out a real case, which could be presented to a compensation board as could be done with so many poisons."

Dr. Alice Hamilton

In 1921 Cecil Drinker's younger brother joined the faculty to direct a course in air analysis. Trained as a chemical engineer, Philip Drinker worked side-by-side with his brother over the next 40 years teaching and developing sampling equipment. This year also marked the

dissolution of the union between Harvard and the Massachusetts Institute of Technology, a requirement of an endowment fund received by Harvard from the Rockefeller Foundation. Unfortunate in some regards, this dissolution allowed the school to accept candidates for studies leading to master and doctor of public health degrees.

Dr. Hamilton focused her teaching on industrial toxicology, and Philip Drinker directed a new program in ventilation and illumination. This program applied engineering principles to measurement of airflow, psychrometry, and design of air conditioning in factories. His courses included direct experimentation with the effects of temperature, humidity, and carbon dioxide; the application of these factors to ventilation

Philip Drinker

efficiency; and the physiological aspects of air conditioning. Animals and human subjects were exposed to dusts, fumes, and smokes to study physiological and pathological effects, and photometric studies were conducted in buildings to study illumination efficiencies.

Constantin Yaglou   Charles Williams   Leslie Silverman

In 1925 Constantin Yaglou joined the department's staff as an engineering instructor, expanding on Drinker's work. He came to the school from the Research Laboratory of the American Society of Heating and Ventilating Engineers, where he researched the effects of temperature, humidity, and air motion on working and resting adults. This research became the basis for the "effective temperature" scale of thermal-comfort and was later modified for premature babies, effectively reducing their mortality. Like Drinker, Yaglou collaborated closely with physicians; thus, forming the basis of modern bioengineering.

Dr. Hamilton retired in 1935 but her contributions continued. Her post-retirement writings have become primary references for the field, including *Industrial Poisons in the US*, *Industrial Toxicology*, and *Exploring the Dangerous Trades*.

In 1937 a geologist joined the Division of Industrial Hygiene as Instructor. Charles R. Williams, having the unique title of "petrographic engineer" while working for Liberty Mutual Insurance Company, researched dusty trades and became an expert on identification and analysis of airborne dust. In 1939 Leslie Silverman, a mechanical engineer and doctoral student was also elevated to instructor. Thus, as World War II began, Philip Drinker, Constantin Yaglou, Charles Williams and Leslie Silverman led the vanguard for industrial hygiene at Harvard.

The war refocused their work, advancing research in pulmonary function, thermal-comfort, and personal protection equipment. A surge of students spurred on by their war experiences entered the school in 1946 with expanded interests in radiological health and air pollution control. The HSPH Annual Report for that year made note that the Harvard School of Public Health was the only graduate school in the US with courses of instruction devoted to industrial hygiene and a staff engaged in research in that field.

Dr. Melvin First, seated in foreground at right

The second half of the century saw great advances in the field of industrial hygiene. Fundamental texts were published (*e.g.*, *Aerosol Technology* by William Hinds and *Ventilation Control for the Work Environment* by William Burgess), and a great deal was learned about the health effects and control of specific environmental pollutants (*e.g.*, Dr. Dade Moeller's research on radon). One constant over the past 50 years has been Dr. Melvin First who has served as a nexus between the pioneers of industrial hygiene and the current environmental health researchers as the Division of Industrial Hygiene expanded into the *Department of Environmental Health*. Dr. First has been involved with designing engineering controls for airborne emissions, indoor and out, and has been instrumental in evaluating biological safety, nuclear air and gas cleaning technology, and the use of ultraviolet to prevent transmission of infectious diseases. Shown to the left is Dr. First with some of his students, many of whom have taken leadership positions in government, industry, health care facilities, consulting firms, and academia over the years.

In the 1970s IEQ research began to expand beyond the traditional focus on industrial hygiene. Greater attention was paid to outdoor air and non-industrial indoor environments. The energy crisis of the 1970s precipitated the use of high-sulphur coal in the US. Scientists and policy makers began to question the scientific basis for outdoor air pollution standards, suspecting that the standards were too lenient. In response, the National Institute of Environmental Health Sciences (NIEHS) was created to conduct research for the Environmental Protection Agency. In 1973 Drs. James Whittenberger, Frank Spiezer, and Benjamin Ferris of the HSPH's Environmental Health Department submitted a proposal that would become one of the most influential and extensive experiments that the school had ever embarked upon to analyse the health effects of air pollution. This study was eventually called the *Six Cities Study*, the goal of which was to explore the relationships between health effects and exposures to outdoor air pollutants in six representative cities across the US.

The results of this study over the past 25 years have provided valuable data on outdoor concentrations of particulate matter, sulphur dioxide, acid aerosols, nitrogen dioxide, and ozone. One of the most important findings of this study, however, was that concentrations of air pollutants indoors were often higher than those outdoors. One of the assistant professors of the time, Dr. John Spengler, and his doctoral students designed personal and home air monitors for the study; this work launched an intensive focus on indoor air pollutants in a wide variety of non-industrial indoor environments including office buildings, medical facilities, vehicles, and homes.

2.2. THE IEQ RESEARCHERS OF TODAY

Dr. Spengler continues to perform research on IEQ. He currently heads the *Environmental Science & Engineering Program* within the Department of Environmental Health. His philosophy remains diverse, and his approach to environmental research is eclectic. As such, he has built a team of researchers from a variety of backgrounds and strives to bridge the gap between the many disciplines he believes necessary to solve IEQ problems. Dr. Spengler and his research team draw upon the traditions of the industrial hygiene pioneers while also searching out innovative methods in order to solve new and old research questions.

Dr. John Spengler

Within the Environmental Science & Engineering Program Dr. Petros Koutrakis leads the development of monitoring equipment and analytical methods to measure indoor, outdoor and personal exposures to air pollutants. He and his fellow researchers have developed a myriad of personal passive monitors, including the Ogawa badges for ammonia and ozone and the Yanagisawa badge for nitrogen dioxide. They have designed the Harvard EPA Annular Denuder System (HEADS) to monitor acid gases and particles by separating gases onto an adsorption and particles onto a filter.

More recently, Professor Koutrakis's research has focused on real-time monitoring with the development of the Continuous Aerosol Mass Monitor (CAMM) that measures hourly particle concentrations and retains the sample for compositional analyses. The equipment developed by his group has been implemented in exposure assessment studies, first exploring ozone, acid aerosols, and sulfates. Evidence from epidemiologic research has focused recent interest on particles. Knowledge gained from earlier exposure assessments, specifically sulfate research, has facilitated an understating of the relationship between ambient and indoor particle concentrations. Recent work has also focused on real-time monitoring of indoor particle size distributions in an effort to identify indoor sources.

Dr. Petros Koutrakis

Epidemiologic evidence has focused recent emphasis on microbiological research within the Program. Dr. Harriet Burge and Dr. Donald Milton head the sampling, analysis, and health effects research of bioaerosols. Dr. Burge's training is in mycology, and her research interests are in allergen characterization, prevalence, and health effects relationships. Her recent projects have included characterization of dust mite, cockroach, mouse, and cat allergens in indoor environments and relating these exposure to health effects in general, particularly

Dr. Harriet Burge

among sensitive populations (*e.g.*, children and asthmatics). She has also investigated aerobiological exposures on aircraft and other modes of transportation. Dr. Burge is studying the ecology of fungal growth on various building materials (*e.g.* ceramic tiles, cellulose insulation, and wall board) and focusing on toxigenic fungi (*e.g.*, *Stachybotrys atra,*) with the aim of describing risks associated with exposure to these fungi. Other work examines the survival of *Mycobacterium tuberculosis* on HEPA filter material. Other large epidemiologic studies, both longitudinal and cross-sectional, examine the relationships between exposure to bioaerosols, other indoor air pollutants, health, comfort and productivity in large office buildings throughout the US. Her research team also investigates health effects of biological aerosols in industrial environments, focusing on Gram-negative bacterial aerosols associated with machining fluids in large metalworking factories.

Dr. Milton's focus is on endotoxins and asthma epidemiology. His current projects include improving sampling methods and assays for endotoxins, exposure assessment of occupational asthma and childhood asthma, and exploring relationships between office building characteristics and employee absence. His research has shown that that personal exposure to airborne endotoxin above 45 Endotoxin Units/m3 was associated with acute decrements in lung

function over the working day. One study is examining whether endotoxin contributes to the onset of wheeze and asthma in a cohort of infants and whether endotoxin in homes can be shown to exacerbate the asthma of their parents and siblings. Another study is examining whether the non-specific building-related symptoms (occasionally referred to as sick building syndrome) is associated with levels of Gram-negative bacteria or endotoxin in office buildings. Recent findings have shown that lower employee absence rates were associated with increased outside air supply and with the absence of humidification. A new study funded by the National Institute for Occupational Safety and Health will reexamine whether these building factors influence employee health

Dr. Donald Milton

and productivity using a blinded intervention design. This study will use Polymerized Chain Reaction (PCR) analysis of nasal lavage fluids to determine the occurrence of respiratory infections among office occupants and whether infection profiles are affected by altering ventilation and humidification in the buildings.

2.3. CURRICULUM

Faculty have defined the essential elements of a broad-based knowledge representing environmental health sciences. As such, there is no formal IEQ program at the School of Public Health. The educational philosophy is to prepare professionals to integrate knowledge across a variety of disciplines, as shown in Figure 1, that will enable them to continue to acquire skills and understanding throughout their lives. Two degrees are offered: the Master of Science and the Doctor of Science. Class work is essential to both degrees with many classes involving aspects of IEQ. These aspects include examples and critical review of past and current research activities. In addition, IEQ sampling methods and analyses developed by Harvard researchers are used in laboratory exercises to illustrate physical and chemical phenomena affecting IEQ. Research in IEQ is another essential component of the curriculum at HSPH. Doctoral students are required to perform original research as part of their degree requirements. Their research often examines in part or in whole issues in IEQ. Of the 37 current doctoral students, approximately half incorporate some aspect of IEQ in their dissertation research. Current dissertations in IEQ cover a variety of topics, including traditional industrial hygiene, environmental epidemiology, indoor source apportionment, animal studies, energy efficiency, and policy.

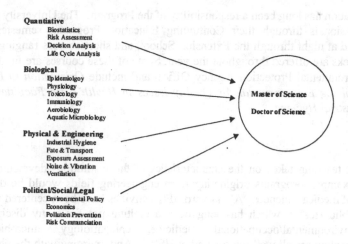

Quantitative
Biostatistics
Risk Assessment
Decision Analysis
Life Cycle Analysis

Biological
Epidemiolgoy
Physiology
Toxicology
Immuniology
Aerobiology
Aquatic Microbiology

Physical & Engineering
Industrial Hygiene
Fate & Transport
Exposure Assessment
Noise & Vibration
Ventilation

Political/Social/Legal
Environmental Policy
Economics
Pollution Prevention
Risk Communciation

Master of Science

Doctor of Science

Figure 1. Environmental Health Courses Offered within the Harvard School of Public Health

Students interested in the indoor environment develop foundational skills within these disciplines and supplement them with courses taken at other schools within the academic community, as shown in Figure 2.

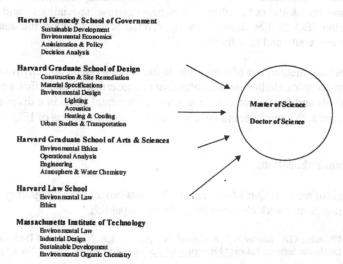

Harvard Kennedy School of Government
Sustainable Development
Environmental Economics
Administration & Policy
Decision Analysis

Harvard Graduate School of Design
Construction & Site Remediation
Material Specifications
Environmental Design
Lighting
Acoustics
Heating & Cooling
Urban Studies & Transportation

Harvard Graduate School of Arts & Sciences
Environmental Ethics
Operational Analysis
Engineering
Atmosphere & Water Chemistry

Harvard Law School
Environmental Law
Ethics

Massachusetts Institute of Technology
Environmental Law
Industrial Design
Sustainable Development
Environmental Organic Chemistry

Master of Science

Doctor of Science

Figure 2. IEQ-Related Courses Provided within the Academic Community

Details of courses are listed on the World Wide Web at:

Harvard University: http://www.harvard.edu
HSPH: http://www.hsph.harvard.edu/registrar
Harvard Continuing Education: http://www.dce.harvard.edu/
MIT: http://web.mit.edu

Professional education has long been a responsibility of the Program. The University offers courses to professionals through their Continuing Education Program.: semester-long courses are offered at night through the Extension School, and short courses ranging from one to several weeks are offered throughout the year. Some of these courses are funded by the regional Environmental Protection Agency Office, and include *Orientation to Indoor Air Quality, Resolving Environmental Air Quality Issues in Health Care Facilities*, and *Principles of Industrial Hygiene*.

3. Conclusions

IEQ research and teaching takes on the characteristics of the institutions developing the programs. For example, programs originating from engineering fields would be distinct from those born of medical sciences. At Harvard, IEQ activities have been centered within its School of Public Health, which has long been a melting pot for many disciplines. Engineering, environmental/occupational medicine, epidemiology, microbiology, environmental sampling are all well represented at HSPH. And, merging with the physical and biological sciences have been the relatively new fields of risk assessment and benefit-cost analysis. Over time, researchers have expanded their interest in the health effects of the manufacturing workplace and urban outdoor environments to environments of homes, transportation systems, offices, health care facilities, and schools where health risks are less severe but affect large populations. The teaching paradigm has remained focused on developing fundamental skills rather than developing narrow specialities, and those interested in pursuing IEQ studies draw on richly diverse schools within the academic communities of the university and its affiliates.

The Harvard model has emerged out of a tradition in industrial hygiene and environmental health to take on the complex challenges of contaminant exposures and stress that a modern society faces. Students and faculty alike must acquire fundamental skills in a diverse array of subject matter in order to contribute effectively in the crosscutting arenas of IEQ.

4. Selection of Recent Publications

Following is a sample of recent IEQ-related research. Students benefit from integrating research with their studies and work closely with the faculty to publish.

Liu, L.J., Olson, M.P., Allen, G.A., Koutrakis, P., Mcdonnell, W.F., Gerrity, T.R., "Evaluation of The Harvard Ozone Passive Sampler On Human Subjects Indoors," Environmental Science & Technology, Vol. 28, No. 5, Pp.915-923, 1994.

Spengler, J.D., Schwab, M., Ryan, P.B., Colome, S., Wilson, A.L., Billick, I., Becker, E., "Personal Exposure To Nitrogen Dioxide In The Los Angeles Basin," Journal Of The Air & Waste Management Association, Vol. 44, Pp. 39-47, 1994.

Reiss, R., Ryan, P.B., Koutrakis, P., "Modeling Ozone Deposition Onto Indoor Residential Surfaces," Environmental Science & Technology, Vol. 28, No. 3, Pp. 504-513, 1994.

Lee, K., Yanagisawa, Y., Spengler, J.D., "Reduction Of Air Pollutant Concentrations In An Indoor Ice-Skating Rink," Environment International, Vol. 20, No. 2, Pp. 191-199, 1994.

Kelsey, K.T., Xia, F., Bodell, W.J., Spengler, J.D., Christiani, D.C., Dockery, D.W., Liber, H.L., "Genotoxicity To Human Cells Induced By Air Particulates Isolated During The Kuwait Oil Fires," Environmental Research, Vol. 64, Pp. 18-25, 1994.

Schwab, M., Mcdermott, A., Spengler, J.D., Samet, J.M., Lambert, W.E., "Seasonal And Yearly Patterns of Indoor Nitrogen Dioxide Levels: Data From Albuquerque, New Mexico," Indoor Air, Vol. 4, Pp. 8-22, 1994.

Brauer, M., Spengler, J.D., "Nitrogen Dioxide Exposures Inside Ice Skating Rinks," American Journal Of Public Health, Vol. 84, No. 3, Pp. 429-433, 1994.

Hornung, R.W., Greife, A.L., Stayner, L.T., Steenland, N.K., Herrick, R.F., Elliott, L.J., Ringenburg, V.L., And Morawetz, J., "Statistical Model For Prediction Of Retrospective Exposure To Ethylene Oxide In An Occupational Mortality Study," American Journal Of Industrial Medicine, 25: 825-836, Wiley-Liss Inc. 1994.

Suh, H.H., Koutrakis, P., Spengler, J.D., "The Relationship Between Airborne Acidity And Ammonia In Indoor Environments," Journal Of Exposure Analysis And Environmental Epidemiology, Vol. 4, No.1,Pp. 1-23, 1994.

Burgess, William A., "Philosophy And Management Of Engineering Control," Patty's Industrial Hygiene And Toxicology, Third Edition, Vol. 3, Chapter 5, Pp. 129-180, 1994.

Sioutas, C., Koutrakis, P., Wolfson, J. M., "Particle Losses In Glass Honeycomb Denuder Samplers," Aerosol Science And Technology, Vol 21, Pp. 137-148, 1994.

Walters, M., Milton, D., Larson, L., And Ford, T.E., "Airborne Environmental Endotoxin: A Cross-Validation Of Sampling And Analysis Techniques," Applied And Environmental Microbiology, Pp-996-1005, American Society For Microbiology, Mar. 1994.

Kramer, M.H.J., And Ford, T.E., "Legionellosis: Ecological Factors Of An Environmentally "New Disease," Zbl. Hyg. 195, 470-482, Gustav Fischer Verlag, 1994.

Lee, K., Yanagisawa, Y., Spengler J.D., And Nakai, S., "Carbon Monoxide And Nitrogen Dioxide Exposures In Indoor Ice Skating Rinks," Journal Of Sports Sciences, 12, 279-283, 1994.

Suh, H. H., Allen, G.A., Aurian-Blajeni, B., Koutrakis, P., And Burton, R. M., "Field Method Comparison For The Characterization Of Acid Aerosols And Gases," Atmospheric Environment, Vol.28, No.18, Pp. 2891-2989, Elsevier Science, 1994.

Sioutas, C., And Koutrakis, P., "Development And Evaluation Of A Low Cutpoint Virtual Impactor," Aerosol Science And Technology 21:223-235, Elsevier Science, 1994.

Sioutas, C., Koutrakis, P., And Burton, R.M., "Development Of A Low Cutpoint Slit Virtual Impactor For Sampling Ambient Fine Particles," J. Aerosol Sci., Vol.25, No.7, Pp.1321-1330, Elsevier Science, 1994.

Leaderer, B.P., Koutrakis, P., Wolfson, Jack, M., And Sullivan, J.R., "Development And Evaluation Of A Passive Sampler To Collect Nitrous Acid And Sulfur Dioxide," Journal Of Exposure Analysis And Environmental Epidemiology, Vol.4, No.4, Pp.503-511, Princeton Scientific Publishing, 1994.

MacIntosh, D.L., Hull, D.A., Brightman, H.S., Yanagisawa, Y., And Ryan, P.B., "A Method For Determining In-Use Efficiency Of Stage Ii Vapor Recovery Systems," Environment International, Vol.20, No.2, Pp.201-207, Elsevier Science, 1994.

Sioutas, C., Koutrakis, P., And Burton, R.M., "A High-Volume Small Cutpoint Virtual Impactor For Separation Of Atmospheric Particulate From Gaseous Pollutants," Particulate Science And Technology, 12:207-221, 1994.

Bellamy, R., Hayes, J. And First, M., "Twenty-Third Doe/Nrc Nuclear Air-Cleaning And Treatment Conference, Design Features," Nuclear Safety, 36(1): 122-134, January-June, 1995.

118

Brauer, M., Dumyahn, T., Spengler, J., Gutschmidt, K., Heinrich, J. And Wichmann, H., "Measurement Of Acidic Aerosol Species In Eastern Europe," Environmental Health Perspectives, 1995.

Brook, J. And Spengler, J., "Exposure To Fine Particle Acidity And Sulfate In 24 North American Communities: The Relationship Between Single Year Observations And Long-Term Exposures," Journal Of The Air & Waste Management Association, 45: 709-721, September 1995.

Burge, H., "Bioaerosols In The Residential Environment," Handbook Of Samples And Sampling, pp.575-593, 1995.

Burgess, W., "Recognition Of Health Hazards In Industry: A Review Of Materials And Processes, (2nd Ed.)," Wiley & Sons, 1995.

Cyrys, J., Gutschmidt, K., Brauer, M., Dumyahn, T., Heinrich, J., Spengler, J. And Wichman, H., "Determination Of Acidic Sulfate Aerosols In Urban Atmospheres In Erfurt (F.R.G.) And Sokolov (Former)," Atmospheric Environment, 29(23): 3545-3557, 1995.

First, M. W., "A Brief History Of Air Cleaning Conferences," Filtration & Separation, 32(5): 449-454, May, 1995.

Hauser, R., Elreedy, S., Hoppin, J. And Christiani, D., "Upper Airway Response In Workers Exposed To Fuel Oil Ash: Nasal Lavage Analysis," Occup Environ Med, 52: 353-358, 1995.

Hauser, R., Elreedy, S., Hoppin, J. And Christiani, D., "Airway Obstruction In Boilermakers Exposed To Fuel Oil Ash," American Journal Of Respiratory And Critical Care Medicine, 152: 1478-1484, 1995.

Lee, K. And Yanagisawa, Y., "Sampler For Measurement Of Alveolar Carbon Monoxide," Environmental Science And Technology, 29(1): 104-107, 1995.

Lee, K., Yanagisawa, Y., Spengler, J. And Billick, I., "Classification Of House Characteristics Based On Indoor Nitrogen Dioxide Concentrations," Environment International, 21(3): 277-282, 1995.

Lee, K. And Yanagisawa, Y., "Sampler For Measurement Of Alveolar Carbon Monoxide," Environmental Science & Technology, 29(1): 104-107, 1995.

Lee, K., Yanagisawa, Y., Spengler, J. And Davis, R., "Assessment Of Precision Of A Passive Sampler By Duplicate Measurements," Environment International, 21(4): 407-412, 1995.

Liu, S., Koutrakis, P., Leech, J. And Broder, I., "Assessment Of Ozone Exposures In The Greater Metropolitan Toronto Area," Journal Of The Air & Waste Management Association, 45: 223-234, 1995.

Neas, L., Dockery, D., Koutrakis, P., Tollerud, D. And Speizer, F., "The Association Of Ambient Air Pollution With Twice Daily Peak Expiratory Flow Rate Measurements In Children," American Journal Of Epidemiology, 141: 111-122, 1995.

Ozkaynak, H., Xue, J., Spengler, J. And Billick, I., "Errors In Estimating Children's Exposures To NO_2 Based On Week-Long Average Indoor NO_2 Measurements," Indoor Air An Integrated Approach, Pp. 43-46, 1995.

Plato, N., Krantz, S., Gustavsson, P., Smith, J. And Westerholm, P., "Fiber Exposure Assessment In The Swedish Rock Wool And Slag Wool Production Industry In 1938 – 1990," Scand J Work Environ Health, 21(5): 345-352, 1995.

Pope, C., Thun, M., Namboodiri, M., Dockery, D., Evans, J., Speizer, F. And Heath, C., "Particulate Air Pollution As A Predictor Of Mortality In A Prospective Study Of Us Adults," Journal Of Respiratory And Critical Care Medicine, 151: 669-674, 1995.

Reiss, R., Koutrakis, P. And Pibbetts, S., "Ozone Reactive Chemistry On Interior Latex Paint," Environmental Science And Technology, 29: 1906-1912, 1995.

Reiss, R., Ryan, P.B., Pibbetts, S. And Koutrakis, P., "Measurement Of Organic Acids, Aldehydes And Ketones In Residential Environments And Their Relationship To Ozone," Journal Of The Air & Waste Management Association, 45:811-822, 1995.

Sioutas, C., Koutrakis, P., Ferguson, S. And Burton, R., "Development And Evaluation Of A Prototype Ambient Particle Concentrator For Inhalation Exposure Studies," Journal Of Inhalation Toxicology, 7: 633-644, 1995.

Sioutas, C., Koutrakis, P. And Burton, R., "A Technique To Expose Animals To Concentrated Fine Ambient Aerosols," Environmental Health Perspectives, 103(2): 172-177, 1995.

Sioutas, C., Koutrakis, P. And Burton, R., "A High-Volume Cutpoint Virtual Impactor For Separation Of Atmospheric Particulate From Gaseous Pollutants," Particulate Science And Technology, 12: 207-221, 1995.

Spengler, J., "Indoor Air Quality - Innovation And Technology," Indoor Air An Integrated Approach, pp. 33-41, 1995.

Spengler, J., Nakai, H., Ozkaynak, H. And Schwab, M., "Housing Factors And Respiratory Health Symptoms: Kanawha Valley, West Virginia," Indoor Air - An Integrated Approach, 1995.

Studnicka, M., Frischer, T., Meinert, R., Studnicka-Benke, A., Hajek, K., Spengler, J. And Neumann, M., "Acidic Particles And Lung Function In Children - A Summer Camp Study In The Austrian Alps," American Journal Of Respiratory And Critical Care Medicine, Pp. 423-430, 1995.

Suh, H., Allen, G., Koutrakis, P., And Burton, R., "Spatial Variation In Acidic Sulfate And Ammonia Concentrations Within Metropolitan Philadelphia," Journal Of The Air & Waste Management Association, 45: 442-452, 1995.

Waldman, J., Koutrakis, P., Allen, G., Thurston, G., Burton, R. And Wilson, W., "Human Exposures To Particle Strong Acidity," Journal Of Inhalation Toxicology, 7: 657-671, 1995.

Delfino, R., Coate, B., Zeiger, R., Seltzer, J., Street, D. And Koutrakis, P., "Daily Asthma Severity In Relation To Personal Ozone Exposure And Outdoor Fungal Spores," American Journal Of Respiratory Critical Care Medicine, 154:633-641, 1996.

Dockery, D., Cunningham, J., Damokosh, A., Neas, L., Spengler, J., Koutrakis, P., Ware, J., Raizenne, M. And Speizer, F., "Health Effects Of Acid Aerosols On North American Children: Respiratory Symptoms," Environmental Health Perspectives, 104(5): 500-505, 1996.

Godish, T. And Spengler, J., "Relationships Between Ventilation And Indoor Air Quality: A Review," Indoor Air, 6: 135-145, 1996.

Gordian, M., Ozkaynak, H., Xue, J., Morris, S. And Spengler, J., "Particulate Air Pollution And Respiratory Disease In Anchorage, Alaska," Environmental Health Perspectives, Journal Of The National Institute Of Environmental Health Sciences, 104(3): 290-297, 1996.

Higashino, H., Tonooka, Y., Yanagisawa, Y. And Ikeda, Y., "Emission Inventory Of Air Pollutants In East Asia (Ii) -Focused On Estimation Of No_2 And Co_2 Emissions In China," J. Jpn. Soc. Atmos. Environ., 31(6): 262-281, 1996.

Lawrence, J. And Koutrakis, P., "Measurement And Speciation Of Gas And Particulate Phase Organic Acidity In An Urban Environment," 1. Analytical Journal Of Geophysical Research, 101(C4): 9159-9169, 1996.

Lawrence, J. And Koutrakis, P., "Measurement And Speciation Of Gas And Particulate Phase Organic Acidity In An Urban Environment," 2. Speciation Journal Of Geophysical Research, 101(C4): 9171-9184, 1996.

Lee, K., Yanagisawa, Y., Spengler, J., Fukumura, Y. And Billick, I., "Classification Of House Characteristics In A Boston Residential Nitrogen Dioxide Characterization Study," Indoor Air, 6: 211-216, 1996.

Liu, L., Burton, R., Wilson, W. And Koutrakis, P., "Comparison Of Aerosol Acidity In Urban And Semi-Rural Environments," Atmospheric Environment, 30(8): 1237-1245, 1996.

Milton, D., Wypij, D., Kriebel, D., Walters, M., Hammond, S. And Evans, J., "Endotoxin Exposure-Response In Fiberglass Manufacturing," American Journal Of Industrial Medicine, 29: 3-13, 1996.

Ozkaynak, H., Xue, J., Zhou, H., Spengler, J. And Thurston, G., "Intercommunity Differences In Acid Aerosol (H+)/Sulfate (So_{24}-) Ratios," Journal Of Exposure Analysis And Environmental Epidemiology, 6(1): 35-55, 1996.

Ozkaynak, H., Xue, J., Spengler, J., Wallace, L., Pellizzari, E. And Jenkins, P., "Personal Exposure To Airborne Particles And Metals: Results From The Particle Team Study In Riverside, California," Journal Of Exposure Analysis And Environmental Epidemiology, 6(1): 57-78, 1996.

Peters, A., Goldstein, I., Beyer, U., Franke, K., Heinrich, J., Dockery, D., Spengler, J. And Wichman, H., "Acute Health Effects Of Exposure To High Levels Of Air Pollution In Eastern Europe," American Journal Of Epidemiology, 144: 570-581, 1996.

Raizenne, M., Neas, L., Damokosh, A., Dockery, D., Spengler, J., Koutrakis, P., Ware, J. And Speizer, F., "Health Effects Of Acid Aerosols On North American Children: Pulmonary Function," Environmental Health Perspectives, 104(5): 506-513, 1996.

Sioutas, C., Wang, P., Ferguson, S., Koutrakis, P. And Mulik, J., "Laboratory And Field Evaluation Of An Improved Glass Honeycomb Denuder/Filter Pack Sampler," Atmospheric Environment, 30(6): 885-895, 1996.

120

Sioutas, C. And Koutrakis, P., "Inertial Separation Of Ultrafine Particles Using A Condensational Growth/Virtual Impaction System," Aerosol Science And Technology, 25: 424-436, 1996.

Spengler, J. And Mccarthy, J., "Indoor Air Quality In Hospitals: Not Just Another Building," Ventilation And Indoor Air Quality In Hospitals, Pp. 3-17, 1996.

Spengler, J., Koutrakis, P., Dockery, D., Raizeene, M. And Speizer, F., "Health Effects Of Acid Aerosols On North American Children: Air Pollution Exposures," Environmental Health Perspectives, 104(5): May 1996.

Spengler, J., "Making Homes Healthier," Indoor Air '96, The 7th International Conference On Indoor Air Quality & Climate, Nagoya, Japan, July 21-31, 1996.

Spengler, J., Dilwali, K. And Samet, J., Indoor Air Pollution, Lesson 26. Pulmonary And Critical Care Update, 11, 1996.

Walters, M., Evans, J., Hammond, S. And Milton, D., "Worker Exposure To Endotoxins And Other Air Contaminants In A Fiberglass Wool Manufacturing Facility," Journal Of The American Industrial Hygiene Association, 1996.

Yoon, D., Lee, K., Yanagisawa, Y., Spengler, J. And Hutchinson, P., "Surveillance Of Indoor Air Quality In Ice Skating Rinks," Environment International, 22(3): 309-314, 1996.

Rao, C., Burge, H.A, And Chang, J., "Review Of Quantitative Standards And Guidelines For Fungi In Indoor Air," Journal Of The Air & Waste Management Association, 46:899-908, Sept., 1996.

Allen, G., Sioutas, C., Koutrakis, P., Reiss, R., Lurman, F. And Roberts, P., "Evaluation Of The Teom' Method For Measurement Of Ambient Particulate Mass In Urban Areas," Journal Of The Air & Waste Management Association, 47: 682-689, June, 1997.

Lazaridis, M. And Koutrakis, P., "Simulation Of Formation And Growth Of Atmospheric Sulfate Particles," American Association For Aerosol Research, 28: 107-199, 1997.

Liu, L., Delfino, R. And Koutrakis, P., "Ozone Exposure Assessment In A Southern California Community," Environmental Health Perspectives, 105: 58-65, 1997.

Robins, T., Seixas, N., Franzblau, A., Abrams, L., Minick, S., Burge, H. And Schork, M., "Acute Respiratory Effects On Workers Exposed To Metalworking Fluid Aerosols In An Automotive Transmission Plant," American Journal Of Industrial Medicine, 31: 510-524, 1997.

Sioutas, C., Ferguson, S., Wolfson, J., Ozkaynak, H. And Koutrakis, P., "Inertial Collection Of Fine Particles Using A High-Volume Rectangular Geometry Conventional Impactor," Journal Of Aerosol Science, 28(6): 1015-1028, 1997.

Sioutas, C., Koutrakis, P., Godleski, J., Ferguson, S., Kim, C. And Burton, R., "Fine Particle Concentrators For Inhalation Exposures -Effect Of Particle Size And Composition," Journal Of Aerosol Science, 28(6): 1057-1071, 1997.

Turpin, B., Saxena, P., Allen, G., Koutrakis, P., Mcmurry, P. And Hildeman, L., "Characterization Of The Southwestern Desert Aerosol, Meadview, Az," Journal Of The Air & Waste Management Association, 47: 344-356, March, 1997.

Verhoeff, A. And Burge, H., "Health Risk Assessment Of Fungi In Home Environments," Annals Of Allergy, Asthma, & Immunology, 78: 544-556, 1997.

Allen, G., Lawrence, J. And Koutrakis, P., "Field Validation Of A Real-Time Method For Aerosol Black Carbon (Aethalometer) And Temporal Patterns Of Summertime Hourly Black Carbon Measurements In Southwestern Pa," Atmospheric Environment, In Press.

Delfino, R., Zeiger, R., Seltzer, J., Matteucci, R., Anderson, P., Street, D. And Koutrakis, P., "The Effect Of The Outdoor Fungal Spore Concentrations On Daily Asthma Severity," Environmental Health Perspectives, 105:622-635, 1997.

Geyh, A., Wolfson, J., Koutrakis, P., Mulik, J. And Avol, E., "Development And Evaluation Of A Small Active Ozone Sampler," Environmental Science And Technology, 31:2326-2330, 1997.

Godleski, J., Sioutas, C., Katler, M. And Koutrakis, P., "Death From Inhalation Of Concentrated Ambient Air Particles In Animal Models Of Pulmonary Disease," Applied Occupational And Environmental Hygiene, In Press.

Ozkaynak, H., Xue, J., Weker, R., Butler, D., Koutrakis, P., And Spengler, J., "The Particle Team (Pteam) Study: Analysis Of The Data," EPA Project Summary, April, 1997.

Wilson, W. And Suh, H., "Fine Particles And Coarse Particles: Concentration Relationships Relevant To Epidemiologic Studies," Journal Of The Air And Waste Management Association, Feb., 1997.

Suh, H.H., Nishioka, Y., Allen, G.A., Koutrakis, P., And Burton, R.M., "The Metropolitan Acid Aerosol Characterization Study: Results From The Summer 1994 Washington, D.C. Field Study," Environmental Health Perspectives, Vol.105, No.8, August, 1997.

Baird, Sjs, Catalano, Pj, Ryan, Lj, And Js Evans. Evaluation Of Effect Profiles: Functional Observational Battery Outcomes. Fundamental And Applied Toxicology, 40: 37-51, 1997.

Ko, G., Burge, H., Muilenberg, M., Rudnick, S. And First, M., "Survival Of Mycobacteria On Hepa Filter Material," Journal Of The American Biological Safety Association, 3(2) Pp.65-78, 1998.

5. Selection of Recent Theses

Following is a sample of recent IEQ-related research conducted by doctoral students for their theses.

1981 Steven D. Colome "Trace Element Characterization of Ambient and Residential Aerosols"

1981 Thomas W. Kalinowski "Aerosol Filtration by a Cocurrent Moving Granular Bed"

1982 Pin-Hua Huang "Determination of 4 MV Bremsstrahlung Spectra by Attenuation Analysis"

1983 George Thurston "A Source Apportionment of Particulate Air Pollution in Metropolitan Boston", "1983"

1983 Ken Sexton "Outdoor, Indoor, and Personal Exposures to Respirable Particles in a Wood-Burning Community"

1984 Janet M. Macher "Evaluation of Airborne Biological Hazards for Occupationally-Exposed Persons"

1985 Edward F. Maher "Control of Radon Decay Product Exposures in Residences"

1985 Judith A. Chow "A Composite Modeling Approach to Assess Air Pollution Source/Receptor Relationship"

1986 Frederic Fahey "The Evaluation of a Gas Scintillation Proportional Chamber for Nuclear Medicine Imaging"

1986 John L. Koehler "Effects of Liquid Utilization on Venturi Scrubber Performance"

1986 Michael R. Flynn "Capture Efficiency of Local Exhaust Ventilation"

1986 Patrick L. Kinney "Assessment of Acute Lung Function Change in Children Exposed to Ambient Ozone"

1987 John F. McCarthy "The Contribution of Environmental Tobacco Smoke to Indoor Particulate Levels"

1987 Robert F.Herrick "Development of a Sampling and Analytical Method for Epoxy-Containing Aerosols"

1988 David P. Harlos "Peak Exposures to Nitrogen Dioxide in Non-Occupational Environments"

1988 John A. Dirgo "Relationships Between Cyclone Dimensions and Performance"

1988 Lorraine M. Conroy "Capture Efficiency of Flanged Slot Hoods"

1989 Lisa Brosseau "Measurement and Prediction of Aerosol Collection by Dust/Mist Respirators"

1989 Carl A. Curling "The Optimization of Filtration for the Reduction of Radon Lung Dose"

1989 Michael Brauer "Human Exposure to Acidic Air Pollutants"

1989 Martin A. Cohen "Assessment of Community Exposures to Volatile Organic Compounds"

1990 Huey-Jen Jenny Su "Health Implications of Airborne Microorganisms in Residential Environments"

1993 Kiyoung Lee "Determinations of Carbon Monoxide Exposure and Dose Using New Samplers"

122

1993 Michael D. Walters "Worker Exposure to Endotoxin and Other Contaminants in Fiberglass Insulation Manufacturing"

1993 Helen H. Suh "Characterization of Acid Aerosol and Gas Exposures in Non-Urban Environments"

1994 Joy Lawrence "Measurement and Speciation of Gas and Particulate Phase Organic Acids"

1994 Man-Sung Yim "Gas-Phase Release of Radionuclides from Low-Level Radioactive Waste Disposal Facilities"

1994 Richard Reiss, Jr. "Ozone Deposition and Reactive Chemistry in Residential Environments"

1994 Lee-Jane Sally Liu "Characterization of Ozone Exposures in Chamber and Field Studies"

1994 Constantinos Sioutas "Development of Virtual Impaction Technologies to Concentrate Fine Ambient Particles"

1995 Kimberly Thompson "Evaluation of National Exposure Assessment Information for Improving Risk Management Decisions"

1996 Shih-Chun Lung "Aqueous-Glass Sorption of Polychlorinated Biphenyl Congeners"

1996 Ho-Yuan Fred Chang "Biological Indicators of 1,3- Butadiene Exposure"

1997 Ginger Chew "Exposure Assessment of Allergens and Culturable Fungi in Residential Environments"

1997 Peng-Yau ErikWang "Continuous Aerosol Mass Measurement by Flow Obstruction"

1997 Youcheng Liu "Assessing Boilermaker' Occupational Exposures to Respiratory Hazards"

1998 Tina Bahadori "Issues in Particulate Matter Exposure Assessment"

1998 Ye Hu "A Chemical Exposure Assessment for a Study of Reproductive Effects in Petrochemical Workers"

PROBLEM BASED TEACHING IN INDOOR AIR SCIENCE AND PRACTICE.

Danish and international experiences

SØREN KJÆRGAARD
Associate Professor
Department of Environmental and Occupational Medicine
Aarhus University
Vennelyst Boulevard 6
DK-8000 Århus C, Denmark

Abstract

This paper discuss experiences with problembased training by a group under the EU-funded ERASMUS program, and experiences at the MPH-program at Aarhus University in Denmark as well as it gives views on the structure and content of a possible common core curriculum, students etc. Problembased training has been applied for long time at Limburg University in Maastricht, The Netherlands. Under the European student exchange (ERASMUS) program an international group of teachers joined to teach a crew of international student group (postgraduates) in Brussels in a week using problembased teaching. The subject of the course was Environment and Health. The program was organized with a few introductory lectures linked to a following group work (with a facilitating tutor), and the next day presentations and discussion among the whole group of students and tutors. The conclusion of the report on the experiment was that the method was useful even in a very mixed group, and highly motivating for the teachers. Similarly, at the MPH-school at Aarhus University, the method has given good experiences in a multidisciplinary group of mature students. Their task have been to evaluate health problems during a month achieving, evaluating literature, and using a strategic algorithm based on 1) problem analysis, 2) setting goals and target groups, 3) selecting intervention, 3) implementation of intervention, and 4) evaluation. An example has been childhood asthma and indoor air pollution. It is suggested that a common core curriculum is constructed as a: 1. Spiraled curriculum, so that students can start at different levels, 2) That subjects is based on a public health point of view, 3) that students should first of all be trained as experts within their discipline, and 4) that training is done as cross-disciplinary group work with facilitating tutors using a problem based training technique.

N. Boschi (ed.), Education and Training in Indoor Air Sciences, 123–128.
© 1999 *Kluwer Academic Publishers. Printed in the Netherlands.*

1. Introduction

The idea of having a common core curriculum among indoor environment scientist and practitioner must be based on the logic of the task they have to perform. That common logic or point of view must be the health of inhabitants or workers living in the indoor environments.

This paper summarizes some views on what types of students to be considered in such an education, the question why a common core curriculum, what structure such a curriculum could have, and how it could be taught in different ways. The base of all the views given here is that students attending such a course do have a basic training within some of the sciences relevant for indoor environment.

2. Students

Considering the field of indoor environment there will always be a need for a multi-disciplinary approach therefore the common core curriculum has to fit a lot of disciplines (chemistry, physics, medicine) and different levels of training like bachelor, master, and doctoral students may have to be taught at different levels, however, one can use the same core curriculum constituents at increasing level of sophistication (a spiraled curriculum). This is also true if one sets up courses aimed at persons with a working experience within the different relevant disciplines. This has been done in the Danish Master of Public Health program with success. At the school we have students with an average age of 40 which are from several disciplines but all have at least 2 years of working experience within public health relevant areas. Students may be nurses, medical doctors, physiotherapists, engineers, administrators (law, political science, social science), dental doctors etc. The consequence of the considerations above is that a common core curriculum somehow becomes the essence of indoor air health and environmental sciences and practice.

3. What is the purpose of a common core curriculum?

The primary purpose should be to learn all scientists and practitioners that they only are experts in a small part of the field, indoor environment and health. For example are medical doctors good a diagnosing and treating diseases but do in general not know enough about epidemiology or about ventilation to make inference about causes in the environment or suggesting preventive measures. Likewise should the ventilation engineer be knowledgeable enough not to give advises on the chemical risks for humans even she have had a small course on toxicology.

So the major objective of training medical doctors, engineers, etc. in the fields of each other, is that they should achieve an understanding to each of them to be able to put out the right questions and to call for help when it is needed.

4. Basic curriculum needs

The goal of 'indoor environmental science and practice' must be the improvement of the indoor environment to fulfill the needs for humans to maintain (or even improve) their health and to be challenged in a way that it improves personal development. This should be achieved in a way that the use of natural resources is minimized and that the health outcome in relation to investments of economical resources is maximized.

Putting this in a strategic thinking model (see figure 1) the needs for a common core curriculum will come out as summarized in the table 1.

Figure 1. The strategic model for prevention of adverse effects in indoor air

Subject	Sub-headings	Keywords
1. Problem description	How many	Descriptive studies Prevalence, incidence
The epidemiology of the disease or symptom	Causes and biology of problem	Analytical (experimental) studies of environmental and individual causes. Toxicology
	Economy	Costs of disease at several levels, e.g. within organization or at society level
2. Problem solution - principal effect of intervention	Effect studies (intervention studies)	Randomized experimental studies of intervention effects on disease or symptoms
Trials of possible solutions	Cost-effectiveness	Economical comparisons of different interventions to find the most cost-effective
	Selection of target groups	Identification of groups susceptible and/or groups exposed to factors in question
3. Goal setting and Target groups	Political decision making based on knowledge of effect	Giving goals to achieve which are possible to evaluate, defining target groups e.g. susceptible groups
4. Implementation	Allocation Controlling output and process for effectivity,	Of resources and man power Monitoring of output and outcome of the program Compliance
	Comparing users and target group	Monitoring of users
5. Comparison with goal and possibly correction of program	Comparing Outcome (monitored health parameter), with goal setting	Either correction of program based on some of the the input from the former step or a return to start

Table 1 Suggestions for content of a common core curriculum

Public health sciences

Statistics and epidemiology	Basic concepts of epidemiology and statistic
Prevention and health promotion strategies	Ventilation technique and strategies, Emission reduction, Building maintenance, Building materials, Building technology, Architecture.
Health economy	Basic concepts of health economy
Communication theory	Presentation of different theories for communication of preventive Measures directed at individual behavior – risk communication.
Risk analysis	Basic concepts of toxicology, Exposure assessment
Risk management	Basic tools to reduce risk

Basic sciences

Physics	Humidity, Temperature, air movements, radiation (radon)
Chemistry	Atmosphere chemistry, Organic compounds Inorganic compounds,
Biology	Microorganisms, Mites, Cockroaches,

Medical and related sciences

Physiology of primary target organs for environmental exposures	External eyes, Mucosa surfaces of nose throat lungs. Nasal functions (odor etc) Lungs functions, Skin functions,
Psychology related to work	Performance Stress
Major diseases and symptoms related to indoor environment	Symptoms, irritation, odor, general etc Allergies Airway diseases Musculo-skeletal symptoms and diseases

5. What training methods should be used.

A basic condition for practicing of interdisciplinary work is that one has to be trained to do it. The most efficient way to be trained is to do it together, so that the students from different disciplines are brought together and trained together. This has for three years been practiced in a newly started Master of Public Health program in Denmark (ref) and with great success. The program is a post-graduate program and recruit students from very different areas as seen from the table. The acceptance criteria for the study are at least a Bachelor and in addition two years of professional work, within fields relevant for public health. The students perform 4 studies on major diseases and their prevention within the strategic algorithm described in figure 1, They participate in group work in 4 groups (5-6 persons) which deliberately are put together with as large variation in professional background as possible. The methodological and factual knowledge about the subject is given through normal lectures and practice sessions, which are related to the actual topic within the area of prevention. During the time the students have to search for literature, do critical reading, condense or meta-analyze, make suggestions and write it all in the form of a report which they later have to present in public for discussion, with other students and tutors.

Table 2 Examples of student background for the Danish Master of Public Health Program

Master level	Bachelor level
Medical doctor	Leading nurse
Dentist	Physio-therapist
Economy	Ergotherapist
Political science	Datalogy engineer
Law	Teacher
Anthropology	
Biology	

With this combination of lectures and own problem orientated work the students both get the theoretical background and learn to work together in a cross-disciplinary way. Later they have to show that they are able to understand and work that way by themselves in the thesis work.

The tasks in the group work have been to evaluate health problems during a month achieving, evaluating literature, in a strategic algorithm basically on 1) problem analysis, 2) setting goals and target groups, 3) selecting intervention, 3) implementation of intervention, and 4) evaluation.

Examples of such group work have been childhood asthma and indoor air pollution, AIDS and health promotion, Cardiovascular diseases and health promotion, Low Back Pain and occupational exposures, Screening for Breast Cancer, How to get smokers to stop, Preventing alcohol and drug abuse.

Our experiences have shown that students with very different background are able to reach a level of high competence and are able to produce qualified theses in the area using scientific methods and thinking.

This method of training has been used for several years in Maastricht at the University of Limburg. Here is problem based training the rule for whole university. The idea of the Maastricht model is to set up the learning goals and set up problems which are able to cover them or part of them so that when students have finished all problem works they have been trough the curriculum. In the indoor air sciences some of the problems could demand medical and epidemiological knowledge in others it could demand knowledge about ventilation and sources. The basic idea is that problems should be formulated in a way so that the students will have to read literature and text books which gives them the needed knowledge, and that all problems are organized in a matrix with problems on one axis and learning goals on the other. Tutors will in both the Danish MPH-program and in the Maastricht model help by facilitating the process and by identifying relevant literature etc.

The Maastricht model have been tried in a week with an international teaching crew and international group of students in a course in Belgium as part of an ERASMUS project. A group of teachers from Belgium (Free University of Brussels), The Netherlands (Limburg University), Denmark (Aarhus University), Sweden (Karolinska University), England (Kings College), Italy (Milan University) joined to teach a crew of international student group (postgraduates) in Brussels in a week using problembased teaching. The subject of the course was Environment and Health, focusing at Air pollution and asthma, Dioxins and breast feeding, Bacterial contamination of drinking water, Cancer Risk in the trucking business, Pesticide guidelines as problems. The program was organized with a few introductory lectures linked to a following group work (with a facilitating tutor), and the next day presentations and discussion among the whole group of students and tutors. The conclusion of the report on the experiment was that the method was useful even in a very mixed group, and highly motivating for the teachers.

The experiences have in both cases been positive from the overall perspective.

6. Conclusions

- Students should have an expertise background of their own field (engineering, medicine etc).
- The common core curriculum should be based on a public health perspective, as this is the overall goal of the efforts.
- The students should be trained together with students from other disciplines.
- This is best done by interdisciplinary group work in carefully planned problem based training, eventually combined with specific lectures and practice within different disciplines.

THE ASSESSMENT TECHNIQUES OF THE INDOOR ENVIRONMENT: THE CASE OF THE ITALIAN UNIVERSITY LIBRARIES

G. GUARNERIO, R. PAVESI
DIPRA, Politecnico di Torino, Italy

Some indoor air quality-related scientific surveys were performed by a group of professors and students through adequate devices, within a graduate course programme. The researchers involved have evaluated parameters deriving from various disciplines: architecture, psychology, physics, chemistry. Consequently, a multidisciplinary approach has been adopted.

The case study was focused on the Faculty of Architecture's Library (Polytechnics of Turin) located in a former XIX century-built laboratory: the impact of the change of use (from the point of view of the indoor environment) was assessed, too. Behavioural considerations, originated from analyses performed according to criteria defined by sociologists, environmental psychologists, and anthropologists, were taken into account, too. The paper aims to describe on the field test methods for evaluating specific indoor enviroments as university libraries are.

The Indoor Environment Quality of a University Library

The educational project results from the contribution of multiple skills from a variety of disciplines. The technical-technological contents, for instance, were supported by the branches of history, law, social-political studies, anthropology, economics, and organisation, etc. This assumption was checked against our training experience in the scope of the Technological Culture Design course held at the Faculty of Architecture of the Turin Polytechnic.

Given the specific skills (researchers and experts), we asked for the contribution of a cultural anthropology expert and for that of the *Environmental System Analysis and Modelling Laboratory* - LAMSA at the *Interdepartmental Training Service Centre of the Faculty of Architecture* - CISDA. A fundamental contribution was offered by the manager and the entire staff of the Library.

Our choice fell on the *Central Library of the Faculty of Architecture* (Turin Polytechnic) at the Valentino Castle, in a refurbished pavilion which previously housed an aeronautic laboratory of the Faculty of Engineering. Currently, the library is part of the Turin

N. Boschi (ed.), Education and Training in Indoor Air Sciences, 129–138.
© 1999 *Kluwer Academic Publishers. Printed in the Netherlands.*

Polytechnic Library System, which co-ordinates all the university libraries, including the Central Libraries of the Faculties of Engineering and Architecture and several sector-oriented libraries. The Library System also co-ordinates all the activities concerning the treatment and circulation of data, which can be consulted at a local workstation or remotely, via Internet, thanks to a computerised system. The wealth of the library is considerable: 60,000 volumes and booklets (50 eighteenth century editions, 1500 nineteenth century editions, 6000 dissertations), 2265 periodicals (of which 505 current), photographs, slides and training material. Furthermore, the public can watch videotapes, read and print microfilms and micro-fiche, as well as use the faculty's internal copy centre.

In 1982, the traditional organisation of the library - where the readers would fill out a form indicating the title and position of the book they were interested in and the material would be distributed by the library staff - was totally changed. An "open shelf" structure was adopted. Readers can choose and take the books without filling out a request form or having to ask the library staff. The library staff is employed in the reading rooms to provide bibliographical information and to tidy the shelves. Before reaching the reading room, the users register at the desk.

This change was welcomed. The direct contact with books is an effective stimulus to increasing knowledge. The long opening hours and the interest for this fundamental place in the faculty - on behalf of both students and researchers - motivated our choice to analyse the issues which concern the library and define its overall "quality".

Indoor environment evaluation tools and methods

CONTEXT EVALUATION MODEL

Architecture alone - notwithstanding its many theories and areas of study - cannot provide the tools required to evaluate the three categories of factors - *technical, functional* and *behavioural* - which connote the environment quality, at the same time. While we have a consolidated research tradition for the first two factors, the users' behaviour description requires human science tools and methods. Environment system specific *technical factors* include dimensions, space typology and topography, hygienic requirements, aeration and lighting requirements, fire escapes and fire-fighting systems. On the other hand, *functional factors* include evaluating the correspondence between the elements and space of the building and their respective usage. These factors represent the configuration and organisation of the physical context in which the activities are performed and on which the quantitative and qualitative satisfaction of the habitation need depends. Consequently, functional indicators include physical-geometrical parameters, i.e. the capacity of a space to house activities and functions. Other issues are: the organisation of individual spaces (usable space, furniture, doors, anthropometric and ergonomic data, flexibility), the configuration of typological units (proximity relationship, position of doors, usable space, rationality, appearance, pleasantness, flexibility, possibility of evolution), the quality of the building ratio with the surrounding environment (accessibility, appearance of the building interface, internal-external relationship, external space design, flexibility, possibility of modification in time), the quality of the facilities (systems and services, technological information, flexibility, possibility of modification in time). Finally, the *behavioural factors* are the variables which describe perception and the ways the space is used: the activity system, movements (obstacles and incentives, origin and destination, voluntary delays and deviations, orientation, conflicting movements) and demarcations (passage traces, use of the space).

BEHAVIOURAL EVALUATION: THE CONTRIBUTION OF HUMAN SCIENCE

A variety of environment evaluation methods and techniques, directed to physical-technical behaviour related issues, were applied in the spirit of this research.
The functional analysis places the space and the user in a system so that the resulting needs lead to a spatial transposition given in terms of performance. The outcome of this type of analysis - generally conducted on a theoretical basis without reference to real contexts and, consequently, without the feature of objectiveness which only field tests considering a certain number of cases can provide - were expressed in the form of graphs and interaction matrixes.

The modality and the specific time requirements of the activities were described with *prossemic methodology* and by means of survey forms referred to the actual context. With respect to the numerous physical, postural, kinetic and individual factors which characterise *personal space* underlying the users' behaviour, the survey specifically dwelled on spatialisation dynamics, closer to the themes treated in architecture.

The path survey in the *Central Library of the Faculty of Architecture* - BCA was conducted using the method of *participating observation* in a prossemic context. Its main objective was to identify the typology and the entity of paths to evaluate functionality, utility and economics in a hypothetical cost-benefit ratio.

Our survey indicated the strict relationship between space and behaviour. Space is considered both as a geometrical surface entity and a time entity, expressed in terms of path speed and stop duration. The survey was conducted on a sample of randomly selected individuals, in different times of the day, in the period from March to May. Before starting the survey, the investigated space was analysed in depth and split into areas according to the activities and the functions performed there.

Methodology and survey techniques

The survey method was based on the use of forms containing the plan of the concerned rooms where the students could mark the paths of the observed individuals with suitable graphical symbols. Other forms were used to mark the most significant behavioural data. Furthermore, the students could add to the general data the features which they considered most explicative for the spatial behaviour of the users. Paths were shown with a line and the slow/fast oppositions of steps and short/long stops were used to describe the type of path and stops of the observed individuals.

CHOOSING A READING TABLE

The surveys conducted on this topic concentrated on the behaviour of the users at the reading tables by observing a random sample of users. The sample included individuals, pairs and groups.

The survey data and the comparison of our data with that of other surveys on the same topic showed that the library is most used in later morning (from 10:30 a.m.) and mid-afternoon (from 3:30 to 7:30 p.m.). The most popular tables were table H and table L (65%) and all those on the loft floor (M and N) with a use percentage ranging from 50 to 90%. The reason for this preference may derive, for the loft floor, by the lower disturbance in the area due to less use and, for the ground floor, by the better environmental conditions, in terms of thermal, acoustic and visual comfort. Table E is located near the door of the ground floor room and is mainly used by groups of students in the afternoon.

Not all the individuals who walk into the library go to a table. Some look for other people, others seem to walk along uncertain paths, consult books standing up and then walk out. In two observed cases, the individuals found a book on the lower floor and took it to a table on the loft floor to read it.

The data on the entity of the previously observed paths was confirmed. The main aisle of the first room on the ground floor is very busy. This is where most of the tables are located. The tables located along the main aisle are generally not preferred by students due to the constant passage of other people. They are commonly used for quick consultation.

By introducing the gender variable, we noticed how the choices of male and female users changed when choosing a table for studying. The degree of possession and privacy, the direction with respect to the doors to control the relationship with others are characterising elements which certainly carry a considerable weight in orienting the choice of the table, when the reduced number of other readers in the library allows.

The use of markers to delimit and defend personal space is actuated by students with the few personal objects which are allowed into the library. The census and the survey on the type, quantity and position of the objects show the need to redefine the standard dimensions assigned for areas where to consult manuals and projects, also considering that large sized drawings are opened when consulting dissertations. Consequently, as readers must leave clothes and bags in a cloakroom at the entrance, the most common personal objects placed on the table to inform strangers of the possession and to avoid unwelcome intrusions are books, magazines, pencil cases, folders, note books, drawing equipment and wallets.

BEHAVIOUR ANALYSIS BY MEANS OF A QUESTIONNAIRE

The space-behaviour relationship was surveyed by processing the data resulting from some questionnaires specifically devised and given to a sample of randomly selected students. The topics chosen and the articulation of the questions varied from questionnaire to questionnaire according to the preference of the interviewers. The questionnaire was put together in two subsequent moments, as the preliminary questions were experimented on the field and demonstrated being ineffective or badly formulated. Some double-check questions were introduced to confirm the validity of the answers.

The results - unlike the participating observations method survey - were represented using statistical techniques and rendered in tables, graphs, histograms and bar charts. The results show that the library is mainly used by female students with respect to male students, prevalently aged between 23 and 25 years old, enrolled in the last years of course, with the following curricula: *protection and restoration* (67%), *architectural design* (22%) and *technology* (11%). As the urban planning curricula has a well equipped and well organised sector library, these students rarely use the *Central Library of the Faculty*.

A relevant number of users (54%) would like furniture more focused on the need for privacy and concentration required for studying, computers at the tables (36%) and drawing tables (54%). 89% of the students deems the service offered by the copy centre inadequate, as the rooms are constantly overcrowded. All the interviewed students would like the library to stay open during the lunch break and more workstations for searches (which often do not give the expected results). Other aspects of dissatisfaction for the current organisation of the library is the time consuming search for books. No-one, in fact, was satisfied by the classification system, also considering that 89% of the interviewees could not remember the codes associated to the topics. Many (89%) think it is important to assign greater space to the horizontal areas and to the reading room, by increasing the number of tables and shelves, deemed insufficient. The absence of lavatories in or near the library raised protests sky-high. Although a certain number of users (between 10% and 30%) claimed their dissatisfaction, the thermal-hygrometry comfort of the rooms (noise, ventilation and temperature) is considered good for a variable number of users ranging from 50 to 81%, notwithstanding considerable differences were noticed in the heating levels and dazzling effects.

EVALUATING INDOOR ENVIRONMENT COMFORT

Environmental wellbeing can be defined as a *mental state*, i.e. an objective state, of neutrality due to the absence of physiological disturbance or as a *state of pleasure*, meaning *the overall conditions providing a feeling of objective pleasure to the individual*. In general, the definition of the qualitative characteristics of an environment tend to privilege the initial definition of comfort - i.e. an objective condition of neutrality. In order to forecast and evaluate the average *sensations* of the users, it is certainly correct to ascertain the absence of physiological disturbance but at the same time to check the actual *measurability* of a mental state and the corresponding scale without disregarding the subjective component of sensations.

THERMAL COMFORT

The condition of *thermal wellbeing* is defined as *the psychological-physical condition of satisfaction which an individual feels for the thermal-hygrometry conditions of the surrounding environment*. Due to the subjective features of this definition, it is in practice replaced by a more objective concept of *thermal neutrality*, condition which occurs when the thermal accumulation of the body is null and the organism keeps both the hot and cold behavioural thermal regulation mechanisms (shivering and sweating) and the vessel-motive thermal regulation mechanisms (vessel constriction and peripheral vessel dilation) nearly inactivate. Thermal neutrality, therefore, depends on the overall environmental conditions - i.e. *microclimate* - conditioning the individual-environment thermal exchanges. Thermal neutrality, in practice, is defined *by measuring physiological and environmental quantities to identify a range of values, rather than a value in itself.*

EVALUATING THERMAL COMFORT

Evaluating the degree of thermal comfort in an indoor environment implies evaluating the quantities which can either be measured directly or parameterised a priori, such as the activity and the clothing of the users, before correlating the data with synthetic indexes - such as PMV (forecasted average) and PPD (forecasted dissatisfied percentage rate).

The PMV index graph analysis - which best expresses the global situation of thermal comfort in the various points of the reading room - shows that a slight sensation of coldness prevails. Point D, located in a reading place on the loft floor, is the only of the four identified points to fall into the comfort range defined in the international ISO 7730 Standard. This may be due to an accentuated stratification of the air temperature higher near the ceiling. In point C, located near a large window on the ground floor, the values fall under the lower PMV limit by -0.5 while points A and B - especially the latter - present a wider range of oscillation. For the latter three points, the number of dissatisfied users - according to the PPD index - is 10% higher than the limit allowed by the standard.

ACOUSTIC COMFORT

Acoustic comfort can be defined as the *psychological-physical condition for which an individual, in the presence of a sound field, declares a situation of satisfaction*. This definition - in mutual relationship with thermal wellbeing - is not suitable without specifying the nature of the acoustic event. Specifically, *noise* is considered as *an unwelcome sound which can cause a unpleasant and annoying acoustic sensation and, consequently, a general state of dissatisfaction towards the acoustic environment*. It is important to note that we will refer to *disturbance* (i.e. alterations which can define acoustic wellbeing) and not *hearing hazards* (i.e. situations in which the hearing capacity can be compromised). This is to distinguish between environments which can be defined *acoustically moderate* (homes, office buildings) and environments where it is possible to express an evaluation more correctly in terms of health hazard rather than comfort (factories, airports, etc.).

EVALUATING ACOUSTIC COMFORT

For evaluating acoustic comfort in the library reading room, we referred to the UNI 8199 Standard. This Standard defines the level of acceptance of the noise produced by heating, air conditioning and ventilation systems (RCV) with reference to the background noise of the environment. In the specific case of the library, the measurements of the background noise and noise output by the running air conditioning system were made in several points in the reading room, considering the areas previously considered for the thermal comfort evaluation, and near some critical areas for the system - such as the air conditioning vents. The measurement were made at 1.20 m from floor level and 0.5 from the walls, as per UNI 8199 Standard.

After measuring the background noise in dB(A) with a one minute samplings in each point, we defined the nature of the noise with the system running in order to evaluate the need to correct the sound level. By recording the equivalent one minute level, the result was a variable nature noise which did not require level correction, as prescribed in the standard.

As regards the acoustic features of the environment, the library presents furniture which does not require experimental measurement of echo time to correct the sound level. Consequently, we considered a conventional echo time of 0.5 seconds. A portable phonometer was used to measure the sound levels.

Noise levels were measured when the system was running and the values were expressed in dB(A). The registered values were higher that the acceptability curve expressed in the UNI 8199 Standard, as shown by the data shown the graph. In particular, the typical area is the most disturbed point by the air conditioning system, exceeding the recommended sound level by approximately 10 dB(A).

VISUAL COMFORT

Illuminating an environment - or rather, more accurately, a visual task - does not only mean directing a certain quantity of light towards it but also means creating a light environment where the users can perform their activities in the most effective and favourable way possible. Only in these conditions can be speak of visual wellbeing. Consequently, the illumination of a closed space must fulfil three functions:
- allow to perform activities and movements in safety;
- allow to perform the visual task in optimal conditions;
- ensure a comfortable internal environment.

EVALUATING VISUAL COMFORT

Visual comfort evaluation regarded both natural and artificial light. The latter, however, was evaluated with the partial presence of daylight, considering it was impossible to make the measurements at night or to totally darken the library windows.

Mainly, two types of measurements were made:
- illumination level;
- luminance distribution.

The illumination measurements were made by means of a portable luxmeter in a series of points on a previously drawn grid in an area of the reading room, identified as typical both for thermal and acoustic comfort evaluations. The mean daylight factor and the illumination uniformity were defined, in the presence of daylight only, from the values measured inside and outside (without direct solar radiation) on the illumination level grid. Similarly, we defined the mean illumination level and illumination uniformity both in the presence of natural and artificial light - a basically constant situation in the library - on the basis of the same grid. Furthermore, we inductively defined the illumination level punctual value for artificial light only, by subtracting the illumination measured in the two previous conditions.

Luminance was measured by means of a lumenmeter with a 1° opening angle. We evaluated the distribution of luminance, in various directions of observation, in several points at the reading tables. The data was then compared to the luminance ratio given in the UNI 10380 Standard. The contrast yield factor was, on the other hand, measured with a contrast yield meter (a lumenmeter fitted on a specific device) in the same points were we measured the luminance distribution. For both types of measurement, we only considered condition of natural light with the contribution of artificial light. We defined the chromatic and colour temperature of light in the typical areas, in the same conditions of luminance, by means of a portable colorimeter.

ANALYSIS OF THE EXPERIMENTAL RESULTS

From the experimental data, the lighting level and the luminance distribution in the library reading room results not fully adequate. The analysis of the values corresponding to the points located on the measurement grid shows a mean illumination level in the typical area sufficient only with the simultaneous presence of the two types of light (natural and artificial) and poor illumination uniformity, especially when the lights are switched off. In actual fact, the natural light is poor inside the library. In fact, the mean daylight factor is 3% lower than the value indicated in the national Standard in force for school buildings. As regards the luminance distribution, the measurements performed in the various reading places (the two maps correspond to a place located in the typical area) show that the lights located on the ground floor of the reading room are not shielded. The ratio between visual task luminance (i.e. a book placed on the reading table) and the luminance of the lights on the observer's visual field is very far from the 1/20 reference value, considering the high values of luminance of the light sources can dazzle. In the considered reading places, on the other hand, the contrast yield factor is adequate. The measurements made in several points (the map corresponds to the measurements made in the typical area) show a CFR value constantly higher than 0.9, thus witnessing the lights are correctly positioned with respect to the reading tables.

As regards the light tonality in the presence of natural and artificial light, we measured a colour temperature equal to 5000 K, corresponding to a neutral tone as prescribed UNI 10380 Standard.

THE QUALITY OF AIR

The issue of indoor air purity, unlike industrial environments, has been underestimated for a long time and resolved with a prescriptive type approach based on defining the ventilation rate exclusively according to the type of environment and the number of people, considering people as the only source of air pollution. Only over the last thirty years have more rigorous studies been conducted on the effects of single pollutants on human health. This lead to outlining a performance approach defining the maximum concentration rate for each single pollutant, as the ASHRAE 62-1989 Standard on building ventilation, which considers both approaches.

EVALUATING AIR QUALITY

We used a dust meter consisting of a hose and fabric filter for the analysis. The air is taken into the device and crosses the filtering element which collects the volatile particles from the air in the mesh. To define the concentration of particulate in the air, we measured the difference in weight before and after intake by means of high-accuracy electronic scales. The mean concentration of particulate results from the weight difference divided by the air volume taken into the meter. The sampling performed in the library reading room lasted 180 minutes and showed a mean particulate concentration under the limit given in the international standard corresponding to 40 mg/m^3 for long term exposure. For a one hour exposure, on the other hand, a concentration of 100 mg/m^3 is acceptable.

THE MULTIMEDIA VISION

Part of our research addressed the creation of a *hypertext multimedia training booklet* for the Central Library of the Faculty of Architecture with computerised tools. *Hypertext* is a collection of *static basic material* in the form of alphanumeric text, graphics and static pictures, created and linked by means of an *author system*. The *multimedia booklet* is an organised set of multimedia training material containing fragments of information - including alphanumeric messages, animated graphics, static pictures and moving pictures - directly visible from the workstation. According to its level, the multimedia product can include sound. All information units are linked and can be accessed interactively by the user by *navigating*, i.e. by selecting and reading the links between the various units. The *links* between topics can be activated by means of graphic buttons or *hotwords* to view digital film footage and graphic animation, listen to audio tracks, create or edit images and graphics, consult databases, etc.

With the *Cyberspace version 1.0* hypertext, we created a sort of handbook for freshman and all the external users who need to acquire spatial behaviour information fast, i.e. the how to use the *Central Library of the Faculty of Architecture* and its paths.

The first phase consisted in collecting, classifying and saving to disk all the information on the Library, hypothesising possible paths which the user can made in the building according to an order which can be changed according to the purpose of the students. The hypertext is articulated in four sections:
- plans
- paths
- rules
- consultation methods

References

UNI EN ISO 7730 (1997) - Ambienti termici moderati. Determinazione degli indici PMV e PPD e specifica delle condizioni di benessere termico
UNI 8199 (1998) - Acustica - Collaudo acustico degli impianti di climatizzazione e ventilazione - Linee guida contrattuali e modalita' di misurazione
UNI 10380 (1994) - Illuminotecnica. Illuminazione di interni con luce artificiale
American Society of Heating, Refrigerating, and Air Conditioning Engineers (1989) Standard 62-1989 *Ventilation for Acceptable Indoor Air Quality*

PART VI. EMERGING ISSUES

A SUSTAINABLE ENVIRONMENT BASIS FOR EDUCATION IN INDOOR AIR SCIENCES

HAL LEVIN
Building Ecology Research Group
Santa Cruz, California, USA

1. What Is Environmental Sustainability?

It has become increasingly clear that human population growth and development activities have combined to place a growing, perhaps excessive burden on environmental resources. Consumption of natural resources, encroachment on land, and emission of pollutants have, together produced strong indicators of environmental stress. Examples of such stress include depletion of the ozone layer and of many natural resources, massive loss of topsoil and of biological diversity, and potential causes of global climate change including increased atmospheric concentrations of carbon dioxide and other greenhouse gases [1-3]. Such environmental stressors generally correlate with economic development and growth [2].

Projected population growth over the next fifty years along with extrapolation of economic development rates suggests a significant increase in environmental burdens [1]. We have used a crude model to estimate an approximate three- to four-fold increase by the year 2050. Others estimate as high as a tenfold increase over the same time period. Regardless of the exact number it is clear that increased population along with increased growth in economic activity will increasingly tax available resources and the pollution assimilation capacity of the earth [1].

The concept of ecological carrying capacity -- or ecocapacity – indicates the extent of resource depletion, pollution, and land encroachment that the earth can tolerate without degradation of the quality of the environment and the services provided by ecological systems [1]. The ecocapacity of the earth translates into the total human population and the per capita share of ecological space or "ecospace" – the amount of the earth's resources and the pollution burdens that can be sustainably supported by the earth..

The shift in awareness and interest in environmental sustainability implies several important themes in future indoor air sciences education. Among these themes is the search for more environmentally benign building materials and building maintenance products. These materials will be more easily cleanable and contain minimally toxic materials compared with dominant materials in use today. Another important theme will be low energy buildings. Such buildings will address issues of illumination, ventilation, heating, cooling, humidity control, and other building equipment using technologies that

141

N. Boschi (ed.), Education and Training in Indoor Air Sciences, 141–150.
© 1999 *Kluwer Academic Publishers. Printed in the Netherlands.*

are more energy conserving and energy efficient. The theme of user control and involvement in building operation will be increasingly represented as a solution to many environmental problems. And finally, recycled buildings will be an important theme in renovation, re-use, and recovery of materials during demolition at the end of a building's useful life [4].

1.1. SUSTAINABLE DEVELOPMENT

Recent years have witnessed a marked increase concern over the relationship between the rates of resource consumption and the pollution emission and the carrying capacity of the earth. This resulted in the establishment of a commission to examine the level of human activity that could be supported by the earth.

1.2. WCED (BRUNTLAND COMMISSION), 1986

The term "sustainable development" was first defined in 1986 by the World Commission on Environment and Development headed by Gro Brundtland, (now head of the Wold Health Organization). The Brundtland Commission defined sustainable development as development that can meet the needs of the present generation without compromising the needs of future generations to meet their own needs [5].

"Meet the needs of the present generation without compromising the needs of future generations to meet their own needs." – WCED, 1986

It's meaning is vague, perhaps intentionally so. Therefore, it has been interpreted differently in diverse contexts and by diverse authorities. Nevertheless, use of the term has increased since that time and some authorities report that there are more than 300 separate definitions of "Sustainable Development.."

1.3. DEFINING SUSTAINABLE DEVELOPMENT (*Dobson: Environmental Politics, 1996*)

A valuable contribution to the discussion of sustainability is the work of Andrew Dobson (who contends that most definitions of sustainability are vague and therefore, not very useful. Dobson tells us that what we mean by sustainability is defined implicitly by the decisions we make on three fundamental questions: what to sustain, why, and for whom? [6].

Dobson has presented an analysis and typology of "sustainabilities" as a means of elucidating a comparative characterization of various definitions and uses of the term. His analysis is summarized in Table 1.

Table 1. Conceptions of Environmental Sustainability [6]

	A	B	C	D
What to sustain?	Total capital (human-made and natural)	Critical natural capital: e.g., 'ecological processes'	Irreversible natural capital	'Units of significance'
Why?	Human welfare (material)	Human welfare (material and aesthetic)	Human welfare (material and aesthetic) and obligations to nature	Obligations to nature
Objects of concern **Primary** **Secondary**	1,3,2,4	1,2,3,4 5,6	(1,5) (2,6) 3,4	(5,1), (6,2) 3,4
Substitutability between human-made and natural capital	Considerable	Not between human-made capital and critical natural capital	Not between human-made capital and irreversible natural capital	Eschews the substitutability debate

Key to numbers:
 1 = present generation human needs
 2 = future generation human needs
 3 = present generation human wants
 4 = future generation human wants
 5 = present generation non-human needs
 6 = future generation non-human needs

Implicit questions of justice:
1. What is to be distributed?
2. Among whom?

Dobson also asserts that there are numerous problems for creation of sustainable projects or communities that are resolved implicitly in one way or another by the choices that are made. These issues are identified in Table Dobson2 along with the solutions presented.

Table 3. Issues which must be resolved in sustainability work. [6]

1	Domain	Cause	Solution
2	Ontological	western science	ontological shift
3	Epistemological	ignorance	precautionary principle
4	Social	gender inequality	women's rights
5		population growth	curbing population
6		lack of property rights in nature	confer property rights in nature
7			
8		debt repayment	debt remission
9		western technology	appropriate technology
10		disempowerment	empowerment
11		poverty	equity
		poverty	wealth
12	Economic	unpriced ecological services	marketised environment
13		trade	protectionism
14		protectionism	trade
15	Institutional	unsustainable resource use	command and control
16		unsustainable resource use	economic instruments
17		international disorganisation	transnational organizations

According to Dobson there are also two types of implicit questions of justice in the resolution of the issues raised in relation to the issues in Table 3: These are 1) Pattern of distribution; and 2.) Procedural or substantive.

1.3.1. *Social and Economic Issues*

Environmental sustainability must be part of a broader strategy and planning process that includes social and economic sustainability as well [2]. Sustainability must be achieved on both economic and social bases as well as environmental ones. The concept of the triangle of sustainability:includes economic, social, and environmental elements.

1.3.2. *Moral And Ethical Issues*

There are fundamental questions of moral and ethical nature that must be addressed, as suggested by Dobson [6]. These include questions of social justice, the distribution of ecospace among rich and poor within and among nations; questions of biological justice, the consideration of species other than humans; and, questions of inter-generational justice, the distribution of resources among today's and future beings, both human and non-human.

2. Buildings Contribution To Environmental Burdens

Buildings are major factors in the consumption of resources and contributors to environmental pollution and land encroachment. These results are due to the enormous amount of material consumed in building construction, maintenance and operation as well as the production of pollution from these processes as well as the consumption of energy for their construction, operation, and ultimate disposal [7]. Figure 1 shows the portion of total human burden on the environment resulting from construction, operation, use, and demolition of buildings in the United States. Results from a survey of global data show similar values.

Table 4. Buildings' share of total human burdens on the environment, data from United States [7]

RESOURCE USE	% OF TOTAL	POLLUTION EMISSION	% OF TOTAL
Raw materials	30	Atmospheric emissions	40
Energy use	42	Water effluents	20
Water use	25	Solid waste	25
Land (in SMSAs)	12	Other releases	13

2.1 PROJECTED FUTURE IMPACTS OF BUILDING DEVELOPMENT ON THE ENVIRONMENT

Impacts of human activities on the environment can be grossly estimated from the combination of population and per capita economic activity. The relationship between environmental impacts and the human actions responsible for them is described by Holdren and Ehrlich in 1971 and has been revisited and revised several times since that time [8]. The expression in equation 1 shows that relationship

$$I = PAT$$
(1)

Where I = impact
P = population
A = affluence (or consumption) per capita
T = the impact per unit of technology

While this general formulation can help us understand the implications of the rates of population and economic growth, it does not help us identify an acceptable target. Specific sustainable targets of impacts must be established for various impacts and in appropriate contexts [9].

2.2. ESTIMATED GROWTH PROJECTIONS

Table 5 presents the project population growth in industrialized and developing countries and global totals for fifty and one-hundred year horizons[9]. It is projected that the majority of growth will occur in the developing nations and that the total increase by the year 2050 will be primarily in the developing nations.

Table 5. United Nations median population projections (billions of people) [3]

Year:	1900	1950	2000	2050	2100	
Industrialized countries	0.6	0.8	1.2	1.3	1.3	Population
Developing countries	1.0	1.7	5.1	8.7	9.9	in billions
World	1.6	2.5	6.3	10.0	11.2	

In Table 6 projected growth in global consumption are projected for the next 50 years. The population growth times the economic growth rates are multiplied to give a crude indicator of projected increases in environmental burdens. Again, the majority of the growth is projected in developing nations. This is due not only to their projected increased population but also their projected continued rapid economic growth. An annual economic growth rate of 2.5% is used for the industrialized nations and 3.5% for the developing nations. This later rate is conservative in light of recent experience [9].

Table 7. Global consumption in 2050 based on population and consumption per capita. [9]

	Now	Population factor	Consumption factor	2050
Industrialized countries	75	1	2	150
Developing countries	25	2	4	200
World	100			350

2.3. ESTABLISHING SUSTAINABILITY TARGETS

Most discussions of sustainability as well as plans and projects claimed to be sustainable are not demonstrably sustainable. Sustainability means or implies the ability to go on indefinitely. Because building designers and builders do not index the environmental performance of their buildings against specific targets calculated on the basis of long-term, indefinite continuation, it is not possible to judge the sustainability of the buildings.

There have been efforts to establish sustainability targets, most notably in the Netherlands. Weterings and Opschoor presented a methodology and calculations to illustrate principles and present results of such calculations [1]. Only when quantitative target values are established and building performance is measured against some apportioned share of total consumption, pollution, and land encroachment can claims of sustainable buildings be justified. Few authors have discussed this question; none have discussed it fully except for the Dutch authors cited above.

Several Swedish researchers mostly centered around Chalmers University argued that the future impacts of human interventions in the environment are impossible to predict reliably. An important illustration is the fact that the ozone hole was only hypothesized and discovered during the past two decades. Prior to that time, humans were unaware of the significance of many pollutants released into the atmosphere. The Swedes, Azar et al, in an effort to find some surrogate or index of sustainability, have produced specific socio-ecological indicators of sustainability [10]. Table 8 shows the fundamental indicators used by Azar and Holmberg and widely promoted by the Natural Step program.. Each of these indicators can be quantified, although some of the indicators for Principle 4 depend heavily on moral and ethical considerations and values.

Table 8. Socio-ecological indicators based on socio-ecological principles [10]

Principle 1: Substances extracted from the lithosphere must not systematically accumulate in the ecosphere	$I_{1.1}$: Lithospheric extraction compared to natural flows $I_{1.2}$: Accumulated lithospheric extraction $I_{1.3}$: Non-renewable energy supply
Principle 2: Society-produced substances must not systematically accumulate in the ecosphere	$I_{2.1}$: Anthropogenic flows compared to natural flows $I_{2.2}$: Long-term implication of emissions of naturally existing substances $I_{2.3}$: Production volumes of persistent chemicals $I_{2.4}$: Long-term implication of emissions of substances that are foreign to nature
Principle 3: The physical conditions for production and diversity within the ecosphere must not systematically be deteriorated	$I_{3.3}$: Transformation of lands $I_{3.2}$: Soil cover $I_{3.3}$: Nutrient balance in soils $I_{3.4}$: Harvesting of funds
Principle 4: The use of resources must be efficient and just with respect to meeting human needs	$I_{3.3}$: Transformation of lands $I_{3.2}$: Soil cover $I_{3.3}$: Nutrient balance in soils $I_{3.4}$: Harvesting of funds

2.4. TRACKING SUSTAINABLE DEVELOPMENT

There are several important activities at the international level that encourage and even mandate sustainable development. Among the most important are the activities of the United Nations Environment Programme and the United Nations Commission on Sustainable Development. The promulgation of Agenda 21 after the Earth Summit in Rio de Janeiro has mandated local, regional, and national Agenda 21 plans. The Conseil International du Bâtiment pour la Recherch l'Étude et la Documentation (CIB) is currently proposing to prepare an Agenda 21 document for Sustainable Construction. Much of the relevant information can be found on the World Wide Web including various sites maintained by the United Nations.

3. Professional Roles In Sustainable Development

There are a number of professionals whose roles in the building environment are critical to a sustainable built environment consistent with good indoor air quality. Each of these professionals and many others will need to be educated to address sustainability issues. These include the following:

- Building designers
- Building constructors
- Building operators
- Indoor environment consultants
- Health and safety officers
- Health care professionals

3.1. LIFE CYCLE ASSESSMENT AND INDOOR AIR SCIENCES

Life cycle assessment (LCA) has become an increasingly common tool to assess the environmental performance of products and materials. More recently, even services are being evaluated using LCA methods and approaches. Only recently have LCAs been performed on building materials and no rigorous LCAs have been performed on total buildings.

LCA practitioners have not included indoor air quality, stating that such inclusion is too complicated [11]. European LCA practitioners have been more aggressive about addressing building materials than their North American counterparts. Furthermore, Europeans have been more willing to produce definitive LCA results indicating products that are preferable from an overall environmental perspective. North American practitioners have tended to limit the use of their LCA results to a set of data without definitive indications of environmental preferability [12].

There is a need to develop analytical tools that will provide assessments of life cycle environmental impacts for total building performance including but not limited to indoor air quality considerations.

3.2. THE FUTURE OF INDOOR AIR SCIENCES IS CLOSELY TIED TO ENVIRONMENTAL SUSTAINABILITY

Economic development issues in the developed and developing countries will define the role of sustainable development. Population growth, while expected to level off in the middle of the 21st Century, will contribute a multiplier to the environmental impacts of economic development in general and building construction, operation and maintenance in particular. As resource scarcity becomes more recognized as a limiting factor on development and well-being, more careful management of both natural and human-made resources will become prevalent. The increased environmental pollution deriving from development and economic growth will raise environmental consciousness throughout the world. More attention to meeting social and economic needs of the less developed nations is a likely impetus for further environmental concern and justice. Finally, political pressures will be brought to bear on all development activity.

3.3. IMPORTANT THEMES IN FUTURE INDOOR AIR SCIENCES EDUCATION BASED ON ENVIRONMENTAL SUSTAINABILITY

The push for environmental sustainability will manifest in buildings both for resource conservation and for protection of human health and comfort. There will be increased demand for buildings made of durable, easily cleanable, minimally toxic materials. These materials will result in less toxic exposure of workers who produce them, install them in buildings, clean and maintain them, and less toxic exposure of building occupants as well. Energy conservation will become an increasingly important goal of sustainable buildings since buildings use around forty percent of all energy used worldwide. Low energy buildings will find new ways to deal with problems of illumination, ventilation, heating, cooling, humidity control. Building equipment will become increasingly energy efficient and have longer useful service lifetimes. There will be increased emphasis on user control and involvement in building operation as a means to improve comfort will minimizing energy and resource consumption. Finally, recycled buildings and building materials will become more dominant. Increased attention will be paid to renovation, re-use, recovery of materials from the existing building stock. Preservation activities will be increased. Recovered building materials will be incorporated into new buildings as well as those that are being renovated or re-constructed.

4. Conclusion

It is clear that environmental sustainability will become an increasingly important societal goal in the 21st Century. For buildings this will include attention to indoor air quality. Consequently, there will be an increased demand for building professionals and building scientists who have been educated in the indoor air sciences as well as in environmental assessment methods. The resulting buildings will be more healthful – that is, they will be less damaging to the environment and more healthful for their occupants.

5. References

1. Weterings, R.A.P.M., J.B. Opschoor (1992). The Ecocapacity as a Challenge to Technological Development. Rijswijk, the Netherlands: Advisory Council for Research on Nature and Environment (RMNO).
2. Daly, Herman (1998) *Beyond Growth*, Boston: Beacon Press Books.
3. Cohen, Joel E., (1995.) "Population Growth and Earth's Human Carrying Capacity." *Science*.269, (July 21) 341-346.
4. Levin, H.(1995) Building Ecology: An Architect's Perspective On Healthy Buildings (Keynote Lecture), In Maroni, M. ed. *Proceedings of Healthy Buildings '95, Volume 1*, Milan, Italy, September 10-14. 5-24.
5. Brundtland, Gro (1986) World Commission on Environment and Development.
6. Dobson, A. (1996). Environmental Sustainabilities: An Analysis and a Typology, *Environmental Politics*, 5 (3), Fall 1966, 401-428.
7. Levin, H, A. Boerstra, and S. Ray (1995) "Scoping U.S. Buildings Inventory Flows and Environmental Impacts in Life Cycle Assessment." Abstract for presentation at Second SETAC World Congress, Vancouver, BC, November 5-9, 1995.
8. Ehrlich, Paul R., Anne H. Ehrlich, John P. Holdren (1977). *Ecoscience: Population, Resources, Environment, Third Edition*. San Francisco: W.H. Freeman and Company.
9. Levin, H. (1997) "Systematic Evaluation and Assessment of Building Environmental Performance," (Keynote lecture) *Proceedings, 2nd International Buildings and Environment Conference*, Paris, France, June 9-12, 1997.
10. Azar, Christian, John Holmberg, and Kristian Lindgren, 1996. Socio-ecological indicators for sustainability. Ecological Economics 18: 89-112.
11. Jönsson, Åsa (1998) Life cycle assessment and indoor environmental assessment, CIB World Congress, Gåvle, Sweden, June 1998.
12. Udo de Haes, H.A. ed. (1996) Towards a methodology for life cycle impact

ENGINEERING EDUCATION FOR INDOOR AIR SPECIALISTS

GEO CLAUSEN, DAVID P. WYON[1]
Centre for Indoor Environment and Energy
Technical University of Denmark

Introduction

Education is about general principles and is intended to equip the student for the future. It should therefore always include a historical perspective, as those who do not learn from past mistakes are bound to repeat them in the future. Training is about passing on current experience and deals with empirical facts rather than theoretical principles, so training is narrow, contemporary and contextual, while education is broad, timeless and universal. The engineer who is too narrowly trained will find it difficult to cope when conditions change or when things go wrong, as they always have and always will. It is easier to train an educated engineer than to retrain a trained engineer whose training has become obsolete. While it may actually be impossible to educate a trained engineer, it is always possible to provide further education for an educated engineer. The purpose of an engineering education is to provide the foundation for lifelong continuous further education and for intermittent training as this becomes appropriate.

New challenges for the engineer

The traditional curricula for the engineer has developed little over the past years. Students were taught to design systems providing a prescribed quantity of outside air to a given indoor volume. The emphasis of the educational system was on technical disciplines such as heat transfer, fluid mechanics, and thermodynamics. New technology has introduced new possibilities for complex calculation, and these new possibilities have led to new engineering tasks. One such example is Computational Fluid Dynamics (CFD). CFD is today a standard tool for the calculation of airflow in buildings and is commonly used by building designers, but this is only possible because of the rapid development and availability of computer power. Although CFD calculation is a revolutionary tool it really consists of solving fundamental equations developed many years ago.

---

[1] David P. Wyon is, besides his affiliation at the Technical University of Denmark, Research Fellow in Indoor Environmental Quality at Johnson Controls Inc., Milwaukee, Wisconsin, USA.

N. Boschi (ed.), Education and Training in Indoor Air Sciences, 151–156.
© 1999 *Kluwer Academic Publishers. Printed in the Netherlands.*

In the future we will see that the same engineer will instead be given the task of providing a prescribed indoor air quality in a given volume. There will be much more focus on human requirements. This change in emphasis calls for a much broader and more multidisciplinary knowledge and understanding of issues such as material emissions, microbial contamination, human senses etc. How much can we reasonably expect of tomorrow's engineer? Naturally, a few lectures on epidemiology does not make an engineer able to solve complex issues within the field of epidemiology, but it will certainly facilitate communication with an epidemiologist in a multidisciplinary team.

It is likely that this transition will take some time as there seems to be a certain reluctance among some engineers to accept this new challenge. This reluctance may be due to uncertainty as to what the implications of this new paradigm are; an uncertainty that could stem from lack of a multidisciplinary education. As the role of engineers is changing, so must the education.

A framework for engineering education in indoor air

There are four educational areas that must be covered. It should be noted that a conventional engineering education comprises only the first and last of these:

Area A1 Building engineering (How buildings are supposed to work)

In the glossy world of building design, air moves decorously in accordance with the laws of CFD, windows, roofs and pipes do not leak, fans are always mounted the right way round, dampers close tightly and open on command, ducts are shiny and tight, condensation takes place in preordained places, cool clear water is added or removed from clean fresh outdoor air and there is no such material as dust. Mould, algae, bacteria, house dust mites, silver fish, rats, discoloured walls, peeling paint and the odour sources associated with these mediaeval aspects of the building trade are therefore impossible and beneath consideration. Light fittings are sources of light and heat and do not flicker, hum, rust, burn dust or fail. Acoustic panels absorb noise and do not harbour dust, house bacteria, shed fibres or smell. Heating elements are sources of radiant and convective heat and do not rattle, bang, leak, rust, heat dust or smell. Building materials have the properties listed in the (glossy) catalogues and do not swell, peel, crumble, rust, oxidise, rot or smell. Occupants are uniform sources of bio-effluent and conform to Normal distributions of response, so their requirements can be represented by a set of equations. If conditions that satisfy 80% of them are provided, it is safe to regard those who complain as a minority and characterise them as deviant trouble-makers suffering from psycho-somatic complaints. The laws of physics therefore rule, chemistry and biology are irrelevant, and Standards may be regarded as written on tablets of stone.

Assuming that this is the case, dr. ing. Pangloss, PhD, PE, is fully justified in teaching HVAC, lighting and noise as separate technical specialities and in leaving the choice of building materials to architects who know very little of all of the above. This "ideal world theory" obviously has a place in the education of engineers who work with indoor air, but it must be presented as what it is: theory, uncontaminated by many of the perturbing factors which together constitute the real world of engineers who work in the building field.

Area A2 Building ecology (The context in which buildings must work)

Buildings, like cars, do not just run on energy, air, raw materials and water - they run on money. They are part of the economy of nations, cities, communities and their individual citizens. It is meaningless to consider building design, operation and maintenance as technical problems without taking account at every stage of the economic context in which these problems must be solved. Building economics is therefore an important part of building ecology. Engineers must be given an understanding of how buildings are financed, and must realise that "building performance" is a term that to real-estate managers denotes return on investment and is measured in monetary units, not in physical outcome variables. They must also understand that buildings are tax objects, so tax law is also a part of the ecology of buildings.

Buildings are major consumers of energy. They are thus subject to all the political and ecological considerations which today govern the use of energy in every form, whether for materials production, transport of materials, construction, operation, demolition or disposal. Life-cycle considerations and sustainable use of resources must obviously be part of an engineer's education today.

Buildings last for many years, so most buildings are not new. Engineers must be given a historical perspective if they are to understand why old buildings were built as they were, how contemporary materials differ from traditional materials, and how new patterns of operation differ from those to which each building was originally adapted when it was designed. Buildings serve the changing needs of society. Understanding the sociological context in which they are designed and used is the only way that an engineer can predict how buildings may come to be used, and thus the only way that buildings can be engineered for the future.

Buildings impact their occupants' lives in ways they will not always accept. The risk of litigation and liability play an increasing role in the economics of building construction and operation. The law of the land as it applies to buildings is part of building ecology. Some knowledge of case law and how many existing buildings each decision might be deemed to cover must become part of an engineer's survival kit in the modern world of building.

Buildings exist in the natural world as well as in the artificial ecologies listed above. Meteorological, geological and biological ways of describing the natural world define the boundary conditions for building and the ecology in which they are operated. Engineers must be educated so that they are able to interface with these interlinked natural and artificial ecologies.

Area A3 Building diagnosis (What usually malfunctions and why)

While all buildings are unique combinations of site, execution and history of use, it is a mistake to regard them in isolation. Buildings throughout history, despite their diversity, have always malfunctioned, while human tolerance has remained very much the same for millions of years. An building engineer's education must include an overview of all the ways that buildings have malfunctioned over time, how, when and by whom this was discovered, who suffered, and what it was necessary to do about it. This will contribute to a healthy skepticism towards any proposal to build the perfect, fault-free building, and will provide the necessary background for the diagnosis of what may have gone wrong with the less than perfect creations they are called upon to operate, rebuild or repair. Cost-benefit calculations should be based on risk analysis over the life cycle of a building, but seldom are. This is why first-cost reduction takes precedence over running-cost reduction. Another reason may be that a building engineer's education places insufficient emphasis on the universal validity of Sod's Law, well known to technical specialists in more advanced fields up to and including rocket science, which states that if something can go wrong, it will, and even if it can't, it may. In the very low-tech, information-starved area of building construction and operation, there is ample proof that this law is well worth bearing in mind.

The avoidable national economic cost of sub-standard building environments in the USA, in terms of BRI, SBS, allergies and asthma, complaints handling, staff turnover and reduced productivity has recently been authoritatively estimated to be $27-168 billion per year, exceeding the cost of avoidance by a factor of 18-47 (Fisk & Rosenfeld 1997). Engineers must be educated to appreciate the cost of malfunction, if they are to be able to argue for design and operation that is cost-effective.

Area A4 Building maintenance (Remediation and prevention)

When buildings malfunction, the last person to know of all those involved is usually the engineer who bears the responsibility for the design. The second last person is the engineer who must do something about it. Building design is essentially an open loop process, and even on-line building operation is "information starved". Engineers must be educated so that they realise the need for both of these engineering roles to be "in the loop", the constant recipient and processor of feedback information of various kinds. This includes physical indicators of building performance - energy use, air change rate, temperature, humidity, noise and lighting levels - but it also includes occupant-related indicators such as sickness absence, BRI (Building Related Illness), SBS (Sick Building Syndrome), occupant satisfaction, complaints and productivity. Occupant-related feedback can provide early warning of physical malfunction even before physical indicators have exceeded permissible levels. This is because permissible levels are a compromise between the often conflicting demands of different occupants. Engineers must be educated to view particularly sensitive occupants, such as those with allergies or asthma, as a source of invaluable and timely information - the canaries in the miner's cage. The communication channels necessary for these loops to be closed must be set up by the engineer, in the form of intermittent occupant satisfaction surveys for maintenance prioritisation, and automated complaint logs linked to service-despatching computer programs and reporting to the responsible engineer.

The need for robust, fail-safe systems with adequate safety margins, systems that allow for human diversity and unpredictability, should be part of a building engineer's education. These are preventive measures. When buildings have malfunctioned, engineers must be aware of the ability of an epidemiological approach to identify the probable cause, and the precision of intervention experiments in proving or disproving cause and effect. Remedial measures which have worked when other buildings have malfunctioned must be taught, together with the need for an exploratory and experimental approach to discovering whether they will work in the currently malfunctioning building.

Examples of new topics in indoor air engineering

CFD; materials emission; airflow patterns; ventilation efficiency; pollutant removal efficiency; purging flow; age of air; tracer gas techniques (for measurement of ventilation rate); passive tracer gas techniques; displacement ventilation; mechanically-assisted or hybrid "natural" ventilation; Demand-Controlled Ventilation (DCV); On-Demand Ventilation Control (ODVC); Task Air Conditioning units (TACs); individual control; raised-floor ventilation; Breathing Zone Filtration (BZF); air cleaners; HEPA filtration; electrostatic filtration; size distribution of airborne particles; respirable particles; allergens; House Dust Mites (HDMs); dust sedimentation rates; resuspension of dust; Surface Dust Contamination (SDC); optimised dust elimination (by ventilation or cleaning); air chemistry; ozone; Perceived Air Quality (PAQ); olf units; decipol units; enthalpy effects on PAQ; Symptom Intensity Feedback Testing (SIFT); the 3.I principle of occupant involvement (Insight, Information, Influence); complaints handling; service-despatching software; automated complaint logs; Occupant Satisfaction Surveys (OS-surveys); Web-based OS-surveys and complaint logs; Occupant productivity; Integrated Facilities Management (IFM); Performance Contracting (guaranteed energy conservation with the necessary investment provided by the guarantor); on-line price optimisation of energy provision from multiple providers..........

Conclusions

The traditional, strictly technical role of engineering in the building field is a trademark of the past. The engineer of tomorrow will face far more diverse challenges involving disciplines that are unlikely to be found in most engineering curricula today. A framework for engineering education in indoor air is suggested: a framework that will equip the engineer for the increasing need for interaction with other specialists which will be a future requirement in the multidisciplinary field of building science.

156

Reference

Fisk WJ, Rosenfeld AH (1997) Estimates of improved productivity and health from better indoor environments. Indoor Air, 7, 158-172

OCCUPANT ASSESSMENT OF INDOOR AIR QUALITY

GARY J RAW, DPHIL CPSYCHOL
Building Research Establishment Ltd, Watford, WD2 7JR, UK

The key aims of the indoor air sciences are to understand how indoor air quality (IAQ) affects the health, comfort and productivity of people in the indoor environment, and to develop and implement methods to improve indoor air quality. This paper first summarises the broad scope of environmental psychology in indoor air science, to place in context the use of occupant assessments. It then explores the role of occupant assessment of IAQ in addressing the above aims. Sick building syndrome is used as an example to show the role and value of occupant surveys, and to introduce some of the relevant methods. All this is intended to represent the basis of what should be included in a curriculum for the indoor air sciences.

1. Environmental Psychology and the Indoor Air Sciences

People are not all the same: different individuals react differently to the same environment. Differences among people should not be seen as causes of symptoms or discomfort but as modifiers: factors that affect an individual's exposure to environmental challenges or response to those challenges. These "individual differences" are due partly to differences in situation (e.g. industrial vs office environment) and physiology (e.g. reactivity to allergenic pollutants) but psychological differences also play their part, in three main ways:

Interpretation
Behavioural response
Linguistic expression.

Interpretation refers to the meaning that people assign to what they perceive. For example, two people might perceive the same odour. One who accepts it as the normal smell of the building will tend to think little of it. The reasoning might be (rightly or wrongly) that it has always been that way and nobody has been harmed by it. Another person, who views the odour as indicating the presence of toxic compounds, will react quite differently, with consequences for personal stress and for the building managers. Such variations may be one of the reasons that visiting panels are such a poor proxy for occupant response (Aizlewood et al 1996): occupants would tend to assign a different meaning to the same stimulus.

N. Boschi (ed.), Education and Training in Indoor Air Sciences, 157–173.
© 1999 *Kluwer Academic Publishers. Printed in the Netherlands.*

Behavioural response refers to the observable responses that occupants make to indoor air quality (IAQ). This would include effects on work efficiency, which would relate to both motivation and capacity to work. Behaviour also includes the adaptive responses of people to environmental challenges: people are not passive recipients of the environment, they adapt by modifying their own exposure (e.g. adding clothing to keep warm), their reactivity (e.g. by taking medication) or their environment, either directly or by requesting others to make changes. Another significant behaviour is the act of making a complaint about the environment or reporting symptoms: this will depend on psychological differences among individuals, probably more than it depends on the actual experience of discomfort or symptoms.

Linguistic expression refers to the way in which people put words to their sensations, beliefs, wishes and behaviours, for example when completing questionnaires about the indoor environment.

Such variations depend partly on factors such as gender, job type and culture but, even within broad groups of people, there are significant individual differences. The response of the individual will itself depend on the response of other building occupants and on hearing about IAQ issues through other routes, such as the media.

The role of management and organisational factors should also not be underestimated. For example, the famous Hawthorne studies (see Roethlisberger & Dickson 1939) showed that environmental change, whether in a positive or negative direction, resulted in improved work productivity. One theory is that the attention given the workers in the study was the causal factor and thus any environmental change might reduce reported symptoms or discomfort, at least in the short term.

To claim that bad management is the cause of IAQ problems can be seen as either claiming an obvious truth (i.e. problems in the workplace are always due to bad management) or propagating an unfair nonsense (i.e. it is not the management, it is the environment in the building). The correct balance between these views can be struck only by establishing in specific terms what management could have done which would have avoided the problems. Broadly speaking, management can be seen as contributing to IAQ problems if it does not act effectively to create a good indoor environment or if it does not establish a good organisational environment for avoiding stress and for dealing with complaints.

One psychological issue, which could clearly be dealt with by better management, is perception of lack of control over the physical environment or the work environment. Lack of personal control, whatever the reasons for it, tends to decrease people's tolerance of discomfort and increase their environmental sensitivity. The notion of control extends to privacy (control over seeing and being seen, hearing and being heard, etc) and being affected by the actions of others (e.g. closing windows, smoking, using chemicals). These aspects of control will obviously depend on the number of people sharing an office, but so does perception of control over heating, ventilation and lighting.

Stress itself is primarily a physiological response but there are individual differences in susceptibility to stress and the actions that follow from stress, including the ability to cope with the stress itself or the external causes of it. Stress can, in turn, increase reactivity to environmental pollutants (Evans 1982).

Apart from differences among individuals, the response of any one individual will vary over time, for similar reasons to those discussed above. This variation might occur over minutes (e.g. because of adaptation to odour or changes in the interpretation of perceptions), hours (e.g. delayed reactions of sensory irritation) or years (e.g. as awareness and understanding of IAQ issues develops). Some psychological variables will be continuously varying over time while other variables will be present or absent or be discrete events. It should also be realised that the selected time period of a study will statistically "censor" the data: if the study is made longer, then more people have a chance, for example, to react with a symptom.

An occupied building functions as a system that includes the interpretations, behaviours and linguistic expressions of the occupants. Each of these factors can have both positive and negative effects on the quality of the indoor environment, and on attempts to assess the effects of the environment upon the occupants. Therefore, researchers and practitioners in indoor air science need to have an appreciation of these issues, for example:

In the process of designing a building or building services;
When monitoring and responding to occupant complaints;
For creating an indoor environment that has a positive effect on productivity.

From the above brief introduction, it is clear that the role of environmental psychology in indoor air science is too broad a field to discuss in detail here. The remainder of this paper therefore deals with one of the more common applications of environmental psychology in the field of IAQ: occupant surveys. While this is only one issue among many, it serves to illustrate many of the principles that must be encompassed when addressing psychological factors in relation to the indoor environment.

2. Occupant Surveys

2.1 THE ROLE OF OCCUPANT SURVEYS

Occupant surveys are widely used to assess the reaction of occupants to the air quality in their indoor environments; such surveys are a powerful tool in both research and practice. However, getting the survey right is not as simple as might at first appear: anyone carrying out or commissioning an IAQ survey should seek to acquire an understanding of the fundamentals described in the following sections.

The starting point for this discussion is to question what we mean by an IAQ problem and how we can know when one exists. The primary concern is the impact of indoor pollution on people, as distinct from effects on building and furnishing materials, animals and plants. In this context, there are three main ways of defining a "problem": by health (identifiable illness or the occurrence of non-specific symptoms), comfort or productivity. Of course, criteria for good IAQ will often find expression in terms of pollutant levels or ventilation rates, but the basis for the criteria is human response. For example, ventilation rates in non-industrial workplaces have generally been set according to criteria of comfort or acceptability since the work of Yaglou et al (1936).

It follows that occupants have a key role in defining the quality of the air in their indoor environments. Apart from the good democratic principle of using people to report their own conditions, the sheer complexity of carrying out measures of indoor pollution points to the value of using people themselves as the instrument.

Objective measurements have the attraction that they should be reproducible and that it should be possible to define precisely what the measurements mean. However, it is not always so clear that they are relevant and it is certainly true that measurements are not perfect predictors of human reaction. It is possible for an investigator to be precisely and accurately measuring the wrong parameters, or measuring the right parameters at the wrong time or place, or with inadequate sensitivity. This is not to suggest that the investigator is incompetent: it is simply a fact that we have an inadequate understanding of how complex mixtures of air pollutants (together with other environmental, social and personal factors) should be measured and how they determine occupant response. The same human responses can result from different IAQ problems, and from factors that are quite distinct from IAQ; therefore it is not feasible to assume that conforming to current IAQ criteria will prevent complaints about IAQ.

Objective IAQ measurements are only as good as their capacity to predict human response; therefore, if human response is available directly, it makes sense to use it. Whatever the instruments say, if the occupants are dissatisfied, there is a problem. It is down to the expert, the practitioner or the consultant to determine exactly what is wrong but it has to be accepted that something is wrong. Conversely, if the occupants were satisfied, it would seem strange to say they should be dissatisfied on the evidence of imperfect measurements. The exception would be where hazardous agents could be present, which the occupants would not be able to detect until it was too late, for example radon or carbon monoxide.

Hence, in assessing IAQ, occupant reactions need to be used alongside environmental measurements (especially if exact identification and/or quantification is required) and medical diagnosis of illness. While there is no sensible case for managing risks of lung cancer or Legionnaires' Disease using surveys of occupant reaction to IAQ, there are many cases in which IAQ problems are best assessed with such surveys. Some information can be acquired only by occupant surveys and by no other means, e.g. symptoms of ill health related to being in a particular building, attitudes to the environment or other factors, and background information on behaviour and factors affecting sensitivity (e.g. allergy). Other information, for example about the environmental conditions in the room, can be acquired in other ways but with the limitations discussed above.

2.2 THE VALUE OF OCCUPANT DATA

In carrying out occupant surveys to investigate IAQ, the investigator can make use of the remarkable capacity of people to act as measuring instruments and data loggers. People can detect, discriminate and describe a wide range of environmental factors, over a huge dynamic range. They can do this for specific factors, locations and times (e.g. the smell of tobacco smoke in the corridor on Monday morning) or combinations of factors, averaged over time and space (e.g. the general air quality in the whole building during the past year). The information can be recorded at the time of exposure or reported retrospectively, without prior warning; this means that people are often the only source of data about how the air quality has been in the past, if a particular environment comes to be blamed for adverse health effects.

Using this rich source of data faces two key problems: poor calibration and inefficient downloading. Calibration refers to the need to use data from different individuals, who respond differently from each other, to make valid judgements about indoor air quality and how to improve it. There are two main ways of overcoming this problem. One is to train people to give similar responses and the other is to average the responses from a large enough number of people that the effects of variation among people are reduced to a level below that which would have an influence on the investigator's conclusions. In some cases, it is possible to accomplish comparability between response scales by transforming the empirical response distributions to the normal distribution (e.g. Lord & Novick 1968) rather than calibrate the scale. Of course, sometimes individual variation is the subject of a study rather than a nuisance to be removed, but this would generally be in a limited number of scientific investigations.

Even if two people have identical reactions to a particular aspect of the environment, they may not give identical linguistic expression in a questionnaire. Not only this, but slightly different questions on the same issue will elicit different answers and it is difficult to predict in advance whether or not a wording change will have an effect (Converse & Presser 1986, Raw et al 1996). This is the problem of downloading data from people. People's breadth of measuring and logging capacity can thus become one of the disadvantages of occupant surveys, if the survey process is not carefully managed so that the investigator knows what the occupants are reporting and according to what criteria. Fortunately, there are procedures for overcoming these problems, as discussed later in this paper.

In conducting a survey, the investigator should be explicit about what is to be measured and for what reason. The principal issues addressed using occupant surveys are health, environmental comfort and effects of IAQ on worker productivity. Surveys can also assess social and personal factors that can modify response to IAQ, such as management issues, personality and sensitivity to air pollutants; however such measurements are outside the scope of this paper.

In comparison with specific illnesses, there is less established knowledge about a range of acute non-specific symptoms which people report when they are in buildings, often referred to as "sick building syndrome". The majority of surveys of occupant reaction to IAQ have been conducted in the context of seeking to explain such symptoms, even when the questions themselves have been about environmental parameters or modifying factors rather than the symptoms themselves. Hence this phenomenon is discussed at greater length below, to illustrate some of the issues surrounding occupant surveys.

It is initially surprising that productivity and staff efficiency have not been more widely used as a direct basis for IAQ standards, but the principle problem has been with defining and measuring productivity. To use work performance as a criterion is feasible in some settings, for example where people are doing repetitive routine tasks, but in other cases it is much more difficult to assess whether performance has been improved or reduced by a certain level of indoor pollution. For some types of work it may be some years after a piece of work was performed before its usefulness can be established. In consequence, there are no readily usable data linking IAQ with productivity. Nevertheless, productivity is a key element in the motivation to improve IAQ in the workplace since it is generally assumed that healthy, comfortable staff are also productive staff.

2.3 THE BUILDING MANAGER'S PERSPECTIVE

Although not normally life-threatening nor disabling, the IAQ problems that can be addressed using occupant surveys (e.g. sick building syndrome) are clearly perceived to be important to those affected, particularly if they are affected at home (e.g. the elderly or sick in residential care) and cannot leave the affected building. Add to this the evidence that the number of people affected and it is clear that the problem is an important one.

Apart from being distressing to the people affected, poor IAQ reduces productivity, increases sickness absence and takes up valuable time in making complaints and dealing with those complaints. Other likely effects are on unofficial time off, reduced overtime and increased staff turnover. In extreme cases buildings may close for a period. If a building gets a reputation for being "sick" it can be difficult to rehabilitate its reputation, even if the building itself is improved. Monitoring to alert management of growing problems is therefore a valuable exercise.

The process of deciding to carry out an occupant survey is important because it should identify the reason for the survey, which in turn will be a key factor in deciding the method to be used. The decision-making process should also bring all the interested parties together in consensus over the approach. The most common reasons for initiating a survey would be:

Because there is a suspected problem with the indoor environment, based on spontaneous complaints, illness reports, sickness absence or environmental monitoring;
Because of a desire to be proactive in monitoring the quality of the indoor environment;
Because of the needs of a research project.

In many cases, the occupants may consider they have already made a diagnosis, that remedial action should be taken and that further surveys are pointless. Such feelings are easy to understand but an unsystematic collection of complaints is a poor indication of the scale and nature of the problems because complaints are encouraged or suppressed by a range of factors unrelated to the subject of the complaint.

At the other extreme, some managers would prefer to believe that there is no problem and surveys will only give staff the excuse they need to complain further. This is unsound. Surveys will identify whether a problem exists and thereby offer the opportunity to solve the problem. To leave the problem hidden is false economy. Either the occupants will continue to be affected, with the implications that has for their health and the efficiency of the organisation occupying the building, or a lot of effort will be wasted in the attempt to cure a problem that never existed.

There is sometimes concern that occupants exaggerate their complaints in order to get something done about their working environment. This however will also be an indication that there is a problem. Again the problem may not be with the environment, it may be with the management, but it still says that there is a problem.

A good occupant survey will do more than confirm (or otherwise) the level of complaint in the building. It will provide information on where and when there are problems, and what types of complaints there are. It may also give an indication of the cause of the problems, but this must always be backed up with independent evidence.

Surveys should ideally be proactive. Carrying out an occupant survey in response to complaints might be called bad management, although it is more conciliatory to call it good management, but late. By analogy, consider a company that did not routinely check the safety of its vehicle fleet but chose instead to wait until a truck got to the point of swerving off the road before checking the whole fleet. If we can be proactive with machines, then why not with people?

By using a standard questionnaire in proactive surveys, the results of the survey can be compared with a wider database to show how the building is performing in relation to the comparable stock. Alternatively, regular repetition of the survey will show whether there are changes over time in the performance of a building, which would give an even clearer indication of impending problems.

The data provided by occupant surveys also serve a purpose in research: to inform those who contribute to future buildings on how to create better environments (whether their contribution is in design, building, installation, commissioning, operation, maintenance or management).

3. Sick Building Syndrome: An Example of the Role of Occupant Surveys

3.1 WHAT IS SICK BUILDING SYNDROME?

The concept of sick building syndrome (SBS) has caused confusion since it was introduced. This section seeks to break through the confusion by offering a usable definition of SBS and showing how this is inextricably linked, through occupant surveys, to the means of diagnosis.

It is inherently difficult to make sense of something if it has no agreed definition that can actually be used in practice. In the case of SBS, there has been a widely-held implicit assumption that something definable exists, while there has been a high degree of variability in the definitions offered; in many studies, no definition at all is given. Within and between nations, there is variation in the words used to describe SBS, but an apparent consistency in the syndrome being described. SBS is defined here as follows.

Sick building syndrome is a phenomenon whereby people experience a range of symptoms when in specific buildings. The symptoms are irritation of the eyes, nose, throat and skin, together with headache, lethargy, irritability and lack of concentration. Although present generally in the population, these symptoms are more prevalent in some buildings than in others, and are reduced in intensity or disappear over time when the afflicted person leaves the building concerned.

Thus, to say that SBS is a real phenomenon is merely to say that there is a variation in symptom prevalence among buildings, not necessarily that there is a single illness called SBS. Neither does the definition imply a clear division into 'sick' and 'healthy' buildings but rather that there is continuous variation.

The cause (or causes) of SBS are at present not clearly identified, but the syndrome can be discriminated from other building-related problems such as:

Complaints about the indoor environment (e.g. temperature, noise, chairs and VDUs);
Long-term cumulative effects of identified indoor hazards (e.g. radon, asbestos);
Specific infectious illnesses caused by known organisms (e.g. Legionnaires' disease);
Building defects which do not cause SBS symptoms (e.g. structural flaws).

Such problems may occur in the same buildings, and the causal factors may overlap, but a distinction is still necessary.

SBS is thus defined, as many health problems have been in the past, in terms of symptoms and conditions of occurrence, rather than cause. The reason is that there is no single proven cause, and any attempt to introduce cause into the definition is likely to be misleading at present, particularly since there are probably multiple causes. It is nevertheless possible to talk of preventive and remedial measures, much as many diseases were to some extent prevented (e.g. by hygiene practices) and treated by reducing specific symptoms (e.g. fever) long before the cause of them was identified.

3.2 THE CAUSES OF SBS

The list of suggested causes for SBS is very long. It is not the purpose of the current paper to rehearse the evidence for various causes but it is worth recalling that, while many studies have focused on IAQ and ventilation rates, there does seem to be some contribution from a wide range of other factors in the environment (particularly temperature, humidity cleanliness of offices and personal control over the environment). Current evidence suggests that no single factor can account for SBS. In all probability there is a different combination of causes in different buildings.

This sometimes results in the statement that we do not know the cause of SBS; this is unhelpful and it is more accurate to say that we know many causes of SBS. The difficulty rests with ascribing a cause in a particular building, since this entails consideration of interactions occurring at a range of levels:

The building: the design, construction and location of a building and its services and furnishings may contribute to IAQ problems in a variety of ways, from the site microclimate through shell design (i.e. depth of space) to the building services and fitting out;

The indoor environment: the effects of the building and site will generally be mediated by the indoor environment;

The occupants: households or organisations which occupy and operate buildings may contribute to IAQ complaints, for example via the quality of building maintenance and work force management;

The individual: reported experience of IAQ problems varies from one person to another within buildings, for a number of reasons which would include personal control over the environment, constitutional factors, behaviour and current mental and physical health.

In addition, action must also be carried through from the original brief, specification and design for a building through the construction, installation and commissioning to the maintenance and operation of the building. Hence it is too simplistic to talk about the causes of SBS only at the level of certain IAQ parameters causing certain symptoms. The implication for occupant surveys is that the investigator must always remember that the determinants of SBS cannot all be addressed in the occupant survey.

3.3 DIAGNOSIS

Although SBS can be defined, the definition of a case of SBS (a "sick building" or "SBS-affected person") is to some extent arbitrary (see Berglund 1990). A theoretical definition of a case could follow from the definition of SBS, but in practice the identification of cases would depend on what is regarded as an acceptable level of symptoms. An analogy would be height: the height of a person can be defined and measured but this does not in itself provide a definition of "a tall person" or "a tall-person building".

In the absence of a meaningful definition of a sick building, there cannot be a valid diagnostic procedure, and thus no clear basis for deciding whether remedial action should be undertaken in a particular building. How then can diagnosis be addressed? Using the definition of SBS suggested in this paper has the implication that the definition is inseparable from the means of diagnosis. This is because the range of symptoms reported in a given building population, and their prevalence, will depend on the number and nature of the questions used to elicit the information. This is, in fact, implicit in any definition of SBS; making it explicit simply brings us closer to dealing with the issue.

Since the definition is tied so closely with the diagnosis, if the means of diagnosis is not agreed, then the definition is also effectively variable between investigations. For example if different questionnaires are used in two buildings, it could be said that two different instruments are being used to measure the same phenomenon. In fact, since what is measured is defined by the questionnaire, two non-identical phenomena are being measured. This becomes most clear when the diagnostic criteria are varied between investigations, for example by using principal components analysis to define syndromes (see Raw 1996).

An attempt to define a working criterion for SBS diagnosis (Raw et al 1990) specifies a level of more than two symptoms per person, experienced more than twice in a year. In a UK cross-sectional survey of 46 buildings (Wilson & Hedge 1987), this criterion fitted 55% of respondents and all the air-conditioned buildings. Such a definition is of any value only if a survey uses the same questionnaire as the one used to produce the original data.

If this appears too academic, of course it can be agreed that it is better to make some attempt to measure SBS than to wait for the perfect diagnostic procedure to be agreed. Nevertheless, the more standardisation there can be in the means of diagnosis, the more we will be able to make sense of SBS. It is important to keep in mind for this purpose that SBS is a complaint of people, not buildings, and can be diagnosed only by assessing the building occupants, not by examining the building itself.

3.4 INSTRUMENTS FOR THE SURVEY

3.4.1 Introduction

Occupant data can be collected by a number of means, including structured or unstructured interviews (medical examinations generally include an interview with the patient and this interview may be more or less structured), discussion groups, and self-completion questionnaires. Although SBS is normally assessed by self-completion questionnaire, this is for convenience and most of the symptoms can be demonstrated by more objective means and shown to be correlated with questionnaire responses. Demonstrations that symptoms can be reduced markedly in blind trials of remedial measures also support the validity of the questionnaires used (Raw et al 1993).

No questionnaire can claim absolute validity as a measure of SBS: the crucial thing is that every time a study is done some key common questions should be used so that it has relative validity. For example, the phrasing of questions can be used to bring about large variations in measured symptom prevalence; such data therefore have little value unless norms are established for a particular standard questionnaire. Norms of close to 50% would tend to give the questionnaire the greatest sensitivity.

In the UK, an expert group set up by the Royal Society of Health has agreed a standard questionnaire, ROES, the Revised Office Environment Survey questionnaire (Raw 1995) together with instructions for use and normative data. ROES is intended to be used for screening surveys to determine the prevalence of SBS in a particular building. The following text is an account of the issues that need to be covered when standardising a questionnaire for SBS, with examples taken from the development of ROES. More detail on ROES itself can be found in Raw (1995) and more detailed general guidance in Berglund et al (1996).

3.4.2 Reliability and validity

One problem in SBS research is that questionnaires have been accepted largely uncritically. It is important to establish the reliability and validity of any instrument, including questionnaires.

The reliability of a measurement refers to how precisely it measures. The reliability is usually expressed as a reliability coefficient which is the proportion of obtained variance that is due to true variance in the variable being measured. Repeatability is one indicator of reliability, since if an instrument is imprecise there will be a low correlation between repeated measurements; this is how reliability was established for ROES.

The validity of a measurement refers to how well it measures what it intends to measure. One type of validity is empirical validity, the degree of association between the measurement and some other observable measurement. ROES was validated in comparison with medical interviews (Burge et al 1990) while others have used objective measures such as tear film breakup (Franck & Skov 1991). An alternative type of validity is construct validity, which means that the measurement should correlate with all other tests with which theory suggests it should correlate, and should not correlate appreciably with other tests with which theory suggests it should not correlate. The ROES symptom prevalences correlate with environmental discomfort and productivity but much less with control over the job.

There is a sense in which subjective reports should not be expected to represent an objective reality, and that they have an implicit validity since it is what the respondent says that is important. This can only be accepted up to a point since, if the report has no relation to physiological states, the investigator will be misled about the nature of the problem

3.4.3 Questionnaire construction

General guidelines have been described by Sudman & Bradburn (1982) and Converse & Presser (1986). There is a need for simplicity, intelligibility, and clarity. It is imperative that common language should be used, questions should be short, and confusions should be avoided (Sheatsley 1983). If the respondent is faced with a task s/he cannot manage or believes s/he cannot manage, the responses have low information value. It is believed that it is easier to answer questions bearing on one's own experience and behaviour ("facts") than questions on opinions and attitudes ("evaluations"). The latter are assumed to be more open to the respondent's own definition than the former. ROES asks about one symptom or discomfort at a time, in short simple questions.

There are generally a number of basic issues to be decided in constructing a questionnaire, including:

Type of response format (e.g. ranks, ratings, magnitude estimation scales);
Open or closed questions (closed questions are easiest to standardise but they are sometimes criticised for limiting the respondent's options);
The effect of the context of other questions, especially neighbouring questions, in the questionnaire;
The overall length and difficulty of the questionnaire (consider the amount of information collected per respondent, how useful the information is, and how many sampled people will respond at all).

The investigators should always understand the measurement instruments to be used in SBS study by:

Piloting or otherwise validating the instrument;
Knowing the meaning of each measure, score or index;
Assessing and correcting for any predictable source of error such as habituation, practice, response sets and false responses.

This may require reading the literature on an instrument (including questionnaires), and being trained in its use. Sometime researchers may rely on developing an understanding of the instrument in the course of the study, but this carries obvious risks.

3.4.4 Specific issues to be considered in SBS questionnaires

Symptom questions can vary in a number of ways:

How many symptoms and which ones (ROES has eight, selected to be representative rather than comprehensive);

Recall period (ROES uses the past 12 months, which would not be expected to give absolutely accurate recall but would pick up problems in whichever season they occur; immediate reporting would be more reliable but produce far fewer responses);

Building-relatedness (ROES seeks to establish whether symptoms are related to being in the target building by asking question "was this better on days away from the office?" but it appears (Raw et al 1996) that the phrasing of the building-relatedness question has little effect on symptom reports);

Response scale (ROES uses a five-point frequency scale with no overlap or gaps but there is a choice of frequency or intensity, numerical or analogue, and the number of points on the scale);

Layout (ROES uses separate questions to achieve fewer missing responses and reduced context effects (Raw et al 1996) but a response matrix takes up less space in a questionnaire).

Ratings of the indoor environment vary in similar ways but seasonal variation will almost always be found and this should therefore be taken into account.

Two kinds of confounding factors can be included in an SBS questionnaire: variables which permit correction of the building symptom score when comparing dissimilar populations (e.g. gender, age) and variables which may provide insight into the reasons for problems in the building. In a screening questionnaire such as ROES, there is only a limited role for seeking to identify the cause of SBS: the existence of the problem should be determined before its causes. Questions about the job and quality of management may inhibit some managers from agreeing to the survey and some staff from returning the questionnaire.

3.5 INSTRUCTIONS FOR QUESTIONNAIRE APPLICATION

A questionnaire is not in itself a method; it is an instrument which will produce valid results if used in accordance with the "manufacturers' instructions". This is of fundamental importance but often overlooked. There is very little value in using a standard questionnaire without following the method prescribed for that questionnaire. The following need to be considered and specialist assistance sought where necessary:

Who should conduct the survey (e.g. an independent organisation, a physician, a personnel manager);

What degree of confidentiality is assured (e.g. questionnaires not associated with individuals; only the investigator knows who completed which questionnaire; completely open);

Survey sampling (e.g. sampling by people vs workstations; universal vs random vs selected groups; if random, simple vs stratified vs multistage);

Survey logistics (e.g. getting approval for the survey, how is survey introduced, how are questionnaires distributed and collected; what account is to be taken of absences from the building, including leave, sickness absence and duty away from study building);
What response rate is required and how should the respondents motivated to achieve the required rate (e.g. management compulsion, rewards, appeal to the value of the survey, written request vs in person);

Analysing and interpreting the data (e.g. how should missing data be treated, how should data be aggregated, what norms are to be used to give meaning to the results, should figures be corrected to take into account the building population features such as balance of male/female);
Which responses can be taken at face value (e.g. reports of dry air can mean that the air is dusty or polluted with organic vapours, ratings of stuffiness can mean that it is too warm and reports of offices which are too warm can be due to low air movement rather than air temperatures in excess of recommended levels)?

3.6 TYPES OF SBS RESEARCH

With a definition and a means of diagnosis (instrument and procedure), consistent research becomes a possibility. But this is only the beginning of making sense of SBS: the design of a study is critical to being able to draw sound conclusions. The difficulty in ascribing causes to SBS can be understood to some extent by looking at the trends over time in the type of research carried out. The history of SBS research can be seen to have passed through three phases, representing a transition from exploratory to confirmatory studies. All three approaches have validity for specific purposes, but each needs to be done well if meaningful results are to be achieved.

The first phase of research effectively commenced in the late 1970s, although there were earlier warnings (Black & Milroy 1966). By the early 1980s, it had demonstrated to the satisfaction of most researchers that there is genuinely a phenomenon which we now call SBS.

Second, notably in the 1980s, there were many investigations that relied on comparisons between buildings. These studies provided evidence on what can be termed 'risk factors': factors which are correlated with SBS but which could not be shown to be causes because of the nature of the studies (e.g. air conditioning, open plan offices and low perceived control over the indoor environment). These factors cannot automatically be regarded as direct causes because of the many confounding factors: it is necessary to discern why each factor is significant. Confounding refers to "a situation in which the effects of two processes are not separated. The distortion of the apparent effect of an exposure on risk brought about by the association with other factors that can influence the outcome" (Last 1983).

Now, in the third phase of SBS research, the "risk factors" constitute important clues as to the causes, clues that are being followed up by making experimental changes to buildings. The basic plan of such studies is first to apply theoretical knowledge and an examination of a building to generate hypotheses about causes of SBS in the particular building being studied. Modifications are then made to the building (and thus to the indoor environment) with concurrent monitoring of both the indoor environment and the reactions of the occupants in order to determine whether the modifications have been successful in reducing the symptoms experienced. This approach controls for most confounders and provides much stronger evidence for the causes of SBS, because most psychological factors and environmental factors remain unchanged. Including a wide range of possible determinants in a study also permits the researcher to identify interactions among causes.

The design of intervention studies has been discussed at length in a recent report (Berglund et al 1996) and recommendations are made on minimum requirements for study design, covering:

Organisation of the study (e.g. use of control groups, double blind design and selection of the study building);
Measurements (e.g. sensitivity to detect changes, validation);
Assessment of outcomes (e.g. specific symptoms vs index or score, time course of changes, maintaining blind assessment);
Assessment of determinants to verify the experimental conditions (e.g. environmental monitoring);
Data analysis (e.g. data validation, within-subject comparisons);
Ethical considerations (e.g. informed consent by building occupants).

3.7 CONCLUSIONS

Making sense of SBS depends not on any single research finding but on putting together the right conceptual framework and using it in research that has been well designed and implemented.

There are necessarily inter-relationships among three important issues regarding SBS:

The definition of SBS;
Diagnosis of SBS in specific buildings;
Establishing and comparing the prevalence of SBS in different buildings and contexts.

This paper has set out a definition which makes diagnosis possible, and a diagnostic method which produced consistent and useful results. The method comprises both a questionnaire and a procedure for using the questionnaire; both are essential. The method is not unique and an indeterminate number of other approaches might be taken. It is better to choose a single approach even if it is for no other reasons than that this particular approach has been tried and has produced a database of comparison figures.

Against this conceptual framework, the benefits and methods of intervention studies have been described.

Future research and problem-solving will need to be directed in an integrated and multi-disciplinary manner to all stages in the life of the building, and to cover the building itself (and its location), the indoor environment, the organisations that occupy buildings and the needs of individual workers. There are many possible causes of SBS and they are interrelated and interactive. SBS is a multifactorial problem which demands a multidisciplinary approach: a comprehensive view and systematic checking of possible problems, not a standard approach applied to all buildings.

References

Aizlewood CE, Oseland NA & Raw GJ (1995) Decipols: should we use them? Indoor & Built Environment, 5, 263-269.

Berglund B. The role of sensory reactions for guides of non-industrial indoor air quality. In Weekes DM & Gammage RB (Eds.), The Practioner's Approach to Indoor Air Quality Investigations. AIHA, 1990, 113-130.

Berglund, B, Jaakkola, J K, Raw, G J & Valbjørn, O. 1996. Sick Building Syndrome: The Design of Intervention Studies. Conseil International du Bâtiment Report, Publication 199. Rotterdam: CIB.

Black F W & Milroy E A. Experience of air-conditioning in offices. J. Institute of Heating and Ventilating Engineers, 1966, 34, 188-96.

Burge, P S, Robertson, A S, & Hedge, A (1990) Indoor Air 90, Vol 1:575-580.

Converse J M & Presser S. Survey Questions. Handcrafting the Standardised Questionnaire. Beverly Hills, CA, Sage, 1986.

Evans G (Ed). Environmental Stress. London, Cambridge University Press, 1982.

Franck C & Skov P. Evaluation of two different questionnaires used for diagnosing ocular manifestations in the sick building syndrome on the basis of an objective index. Indoor Air, 1991, 1, 5-11.

Last J M (Ed.). 1983. A Dictionary of Epidemiology. Oxford: University Press.

Lord F M & Novick M R. Statistical Theories of Mental Test Scores. Reading, MA, Addison-Wesley Publishing, 1968.

Raw, GJ (Ed) (1995) A questionnaire for studies of sick building syndrome. BRE Report. London: Construction Research Communications.

Raw, G J. 1996. Lies, damned lies, statistics and principal components analysis. Indoor Air 96, Vol 1:453-458. Nagoya, Japan.

Raw, G J, Roys, M S & Leaman, A. 1990. Further findings from the office environment survey: productivity. Indoor Air 90, Vol 1:231-236.

G J Raw, M S Roys & C Whitehead (1993) Sick building syndrome: cleanliness is next to healthiness. Indoor Air, 3, 237-245.

Raw G J, Roys M S, Whitehead C & Tong D. 1996. Questionnaire design for sick building syndrome: an empirical comparison of options. Environment International, 22:61-72.

Roethlisberger F S & Dickson W J. Management and the Worker. Cambridge, MA, Harvard University Press, 1939.

Sheatsley P B. Questionnaire construction and item writing. In Rossi P H, Wright J D & Anderson A B (Eds.), Handbook of Survey Research. New York, Academic Press, 1983.

Sudman S & Bradburn N M. Asking Questions: A Practical Guide to Questionnaire Design. San Francisco, CA, Jossey-Bass, 1982.

S Wilson & A Hedge (1987). The Office Environment Survey. A study of Building Sickness. London: Building Use Studies.

Yaglou, CP, Riley, EC & Coggins, DI (1936) Ventilation requirements. ASHVE Transactions, 42, 133-162.

KNOWLEDGE AND TECHNOLOGY TRANSFER IN TEACHING IN INDOOR AIR SCIENCES

L. MORAWSKA
Queensland University of Technology
Brisbane, Australia

The focus of the paper as presented in the title can be expressed in an alternative way, which is as a question: how universal are approaches to teaching and training in the area of science and practice of indoor air quality? And also: is it possible, practical or desirable to transfer training or university degree programs from one geographical, cultural or economical reality to another? The paper is an attempt to answer the two questions from the broad perspective of linking parallels between teaching in indoor air sciences and teaching in an interdisciplinary area in general, and also from the perspective of personal experience from university and training course teaching in various places in the world.

As a starting point for this discussion the issue of how universal any teaching programs are, could be investigated. For this purpose a brief analysis of a few different areas of university teaching will be discussed. Let us look first, for example at a university degree in physical science. Would there be any difference in the program if it was taught in New York, Budapest or Kuala Lumpur? I cannot see any reasons for differences in the core units of the program. The only difference could relate to some elective subjects, which are often outside the discipline of study. The same would apply for example to chemistry. What about engineering? The vast majority of the program would be the same, however there would be differences relating to different standards and norms used in different countries, or materials applied in different climatic zones. Law programs? In most cases the fraction of units that are locally specific would be much higher than in engineering. So in summary, while certain single discipline university teaching programs are universal, some would differ too significantly to be considered transferable.

Teaching and training in science and the practice of indoor air presents a much more complicated case, as this is not a single discipline, but a very interdisciplinary and very applied area. It is thus unavoidable that while certain aspects of the teaching and training programs are universal, other aspects are country or area specific. How large a problem does it present, and what are the solutions for the exchange of programs?

N. Boschi (ed.), Education and Training in Indoor Air Sciences, 175–180.
© 1999 *Kluwer Academic Publishers. Printed in the Netherlands.*

Universal versus local approaches to education and training in IAQ Similarities and differences between different levels of education and training in IAQ can be considered in three key areas:

- Content of the material
- Organization of teaching and training
- Economical aspects: possibilities and priorities

Content of the material

In considering teaching and training programs in indoor air science and practice a key point is that the area of study is interdisciplinary. While there is an ongoing discussion on defining whether indoor air science is an independent discipline or not, I will take for the purpose of this paper that indoor air science is not itself a discipline, but applies knowledge from many disciplines to the analysis and solution of indoor air and indoor environment problems (this is a modified definition of environmental science). Thus, at the completion of an education or training program in this area, the student (graduate) should be equipped in expertise in a specific discipline, as well as in the broad field of indoor air science. I will introduce here a concept of a three level teaching structure in application to teaching in indoor air sciences.

A very important aspect of teaching in the area of indoor air sciences is a systems approach: the indoor environment has to be seen as a system, and any investigations or control techniques have to be approached first from a system level, before isolating a specific component of the system for more detailed, often unidisciplinary considerations. This is a principal requirement of teaching and training in this area, and as such is universal. It will be considered here to constitute the first level of the teaching structure. Another universal aspect is the mix of disciplines required for a comprehensive approach to this area. Indoor air and environment experts are recruited from those whose primary education background is in science (physics, chemistry, biology, etc.), engineering (mechanical, ventilation, building, etc.), health (physicians, nurses, occupational practitioners, etc.), management, law, etc. A complete understanding of the systems approach requires understanding of the roles and contributions of individual expertise areas to the system. This will be considered here as the second level of the teaching structure.

The differences in the content of the material would become more pronounced at the third level of this teaching structure, which is the level of individual disciplines. But, this relates to the conclusion drawn above that while certain single discipline teaching programs are universal, some differ too significantly to be considered transferable. Since this, however, is the third level of the teaching structure in this area, most of the program could be considered transferable.

There should be a differentiation here, between larger, more general programs (such as university postgraduate degrees), and often much more narrowly oriented training programs. Since the university programs would encompass all three levels of the teaching structure, they would be much more transferable, according to the conclusion above, than some of the training programs, which relate to the third level of this structure.

Organization of teaching and training

The key issues to consider in this area are:

- who are the target students,
- what is the purpose of learning (for example: know how to comply with the regulations, broadening of knowledge, or professional requirements),
- local requirements for teaching and training and local organization of teaching.

The first two dot points relate more to training, while the third one to both training and teaching. The universal aspect here is, that target students and the purpose of learning have to be the same in the places (countries, regions) where the training is to be offered compared to where it was developed, to consider the program transferable. For example if a particular, narrowly oriented training course is designed for example, for government managers, in a particular country, to learn how to use and apply a particular software package to estimate exposure risk for the purpose of compliance with regulations, there will be little value in offering this course in a different country. Different regulations and different procedures for compliance with regulations make training of this nature applicable only to this country. While there are many examples of unification and harmonization in regulations and requirements resulting in unification of training programs(an example a training program for European Commercial Environment Expert), it is not feasible that in any foreseeable future world wide or even regional unification will be achieved.

Organizational aspects of teaching and training often relate to what is required after completion of the program. A degree (with, for example, a required distribution of credit points), an exam passed, a certificate of attendance, are examples of the requirements that will allow a person graduating from a program to be considered as fulfilling local specific professional criteria. While these aspects are usually of a more administrative and organizational nature, they could create real obstacles in program transfer. For example, accreditation of a course or a degree program offered by one university, to be offered at another university, is usually a task which takes many month if not years.

Economical aspects: possibilities and priorities

This aspect could be a key element, deciding not only that a training or teaching program on the quality of indoor air and environment could be conducted and how, but more fundamentally, whether this is considered an area of concern or not. In many developing countries where there is shortage of food, water, shelter, medical care, the quality of the indoor air would be at a very low level of priorities. It is obvious that it would not be a point in transferring a training course on maintenance of filtration and ventilation systems, if we talk of an indoor environment as a tent or thatched hut. An irony here is, that technological solutions to significantly improve the quality of air in such environments exist and often are very simple. But on one hand there is no basic awareness that this could be a problem, and on the other hand the cost of this solutions (for example a properly vented stove), which would be insignificant in developed countries, would be totally prohibiting in developing countries.

But even in developed countries, where the cost would not be a major problem, it would not be a priority to ensure good quality of indoor environment if there are no regulations (standards and guidelines), which would make it mandatory to address the issues of the quality of the indoor environment. In both these cases, the key aspects of training would be not to focus on what to do or how to do it, but on the need to do it.

Solutions for exchange of education and training programs

Most of the how to do it type education and training programs could be considered transferable between countries and regions, unless the focus of a program is so narrowly linked to some specific aspects of the local regulations or approaches that it is not applicable outside the country or jurisdiction boundaries. The need to do it type programs would not be considered as transferable, but as developed by the program designer for a specific area or country.

The most important aspect of what to do or how to do it programs is the system approach to learning, and the more general the program is, the easier it is to implement this approach. Thus in general, transfer of university programs would present mainly organizational and administrative problems, while transfer of more narrowly oriented courses would present more problems, ranging from the need for content modification to considering the program non-transferable.

A very important aspect of general type programs, teaching a systems approach to indoor air, is to ensure that students obtain a very strong background in one specific area of knowledge or expertise, and a broad, umbrella type knowledge in relation to the whole area. This approach is not unique to teaching in indoor air science and practice, but to any interdisciplinary teaching. An example of this is the environmental science major, undergraduate program developed and offered by the Queensland University of Technology, Brisbane, Australia. To summarize the program, undergraduate students undertaking the environmental science major, are strongly recommended to undertake this either as a double major or major and minor in any of the following areas: mathematics, physics, chemistry, earth sciences or life sciences.

The environmental science major program has four units which are the same for all environmental science students, while four other units are closely related to and strengthen the areas which are of direct significance for environmental applications (for example an extended program in fluid mechanics and transport theory for environmental science major/physics major students) (Morawska, 1998). Another important aspect of interdisciplinary programs is, that the systems approach students learn is broad enough, that it could be applied to systems and situations, other than those which were the main focal areas for the students.

Important global aspects of education and training in the area of indoor air sciences, discussed extensively above, are clearly defined objectives and target students, for the program. Programs which clearly identify both aspects, are much easier to transfer, if the clearly defined requirements on the receiving end are similar.

How to address the differences, of more narrowly oriented programs, such as training courses? The best recipe for courses like this is:

- the systems approach component and global aspects of the program can be taught by the program designer or provider,
- area or country specific aspects should be taught by a local expert or in collaboration with a local expert.

Examples of these approaches are Indoor Air Quality Assessment courses, in Sydney, Australia of which about two thirds were taught by a Canadian expert and one third by Australian experts. The component presented by the Australian presenters consisted mainly of case studies related to the Australian climatic conditions and regulation compliance requirements. Another example is a training course: Indoor Air in Hospital Environment, presented by the author of this paper to a large audience of engineering and medical practitioners from health care facilities in Taipei, Taiwan. About three quarters of the course was the author's presentation on various general and hospital specific aspects of indoor air, with a focus on particle pollution. The remaining quarter was presented by Taiwanese researchers and practitioners, and consisted of a report from a project conducted in several hospitals in Taiwan on air quality in hospital isolation rooms. Another example is a broader training course offered by several experts from different countries and continents, representing the International Society of Indoor Air Quality and Climate (ISIAQ). The course has already been presented in Sweden and the USA (and included Sweden and USA focused presentations), and attended by those whose professional duties require a broad knowledge of global trends and approaches to indoor air quality and environment. In all three cases the courses received top assessments by the participants.

In conclusion, transfer of knowledge and technology in teaching in the area of indoor air sciences is possible and desirable. The success of program transfer depends on the understanding of the philosophy of teaching in this area as well as on understanding local needs, requirements and limitations. The globalisation of education and training require continuous upgrades of the teaching programs to follow technology developments, as well as the ever changing social and economical conditions and requirements of countries and regions.

180

References

Morawska L, Interdisciplinary Teaching at the Undergraduate Level: Queensland University of technology Environmental Science Program. Submitted for publication.

PART VII. DIFFERENT EDUCATIONAL PLAYERS FOR DIFFERENT AUDIENCES AND NEEDS

FEDERAL ROLES IN EDUCATION AND TRAINING IN THE INDOOR AIR SCIENCES

Case Studies of U.S. Environmental Protection Agency Involvement in Training Development and Delivery

ELISSA FELDMAN
Associate Director, Indoor Environments Division
U.S. Environmental Protection Agency
401 M Street, S.W.
Washington, D.C. 20460

1. Introduction

The U.S. Environmental Protection Agency's (EPA) Indoor Environments Division, which prior to 1995 was two distinct Divisions, the Radon Division and the Indoor Air Division, has played several different roles in the development and dissemination of education and training courses in the indoor air sciences. In this paper, three case studies are presented which illustrate these differing roles: Government as developer and disseminator of a training course; Government as developer of a training course and facilitator of its dissemination by others; and Government as catalyst by which others develop and disseminate a training course.

2. Case Study: EPA's Radon Proficiency Programs

EPA initiated the Radon Action Program in 1985 as its first response to a national public health threat presented by radon contamination of indoor air in homes. A key activity of the program has been to foster State and private capabilities to reduce radon risk. With passage of the Indoor Radon Abatement Act in 1988, EPA has made funds available annually to States to develop and operate radon exposure reduction programs, and has provided training courses for companies and individuals entering the radon measurement and mitigation field. In addition, since 1989, EPA has operated a voluntary mitigation and measurement proficiency program for companies and individuals who wish to provide radon services, offering examinations for providers who, upon successful performance, have been listed by EPA as proficient. The list has been made available to the public so consumers can obtain services from competent measurement and mitigation providers. This year (1998), responsibility for operating the proficiency program is passing to a private entity, although EPA will maintain oversight responsibilities [1].

N. Boschi (ed.), Education and Training in Indoor Air Sciences, 183–189.
© 1999 *Kluwer Academic Publishers. Printed in the Netherlands.*

The Radon Training Program evolved over the course of several years. In the early years from 1985-1987, EPA developed a base of knowledge on radon mitigation and transferred this information to States and the private sector through a basic training course and publications. Because radon was a new public health issue, a broad range of audiences, including State and local government officials, the private sector, and interested homeowners had to be educated to understand the nature of indoor radon and its health risks. In addition, specific training in radon mitigation was needed to address homes where elevated radon levels were identified. This presented a challenge to preparing a comprehensive training course which could encompass the breadth of backgrounds of people likely to be interested in learning about the issue.

In 1985, EPA and the New York State Energy Office began compiling and organizing information regarding all aspects of indoor radon. These materials evolved into a comprehensive reference and instructional training manual, A Reducing Radon In Structures. The manual and supporting audio-visual materials became the foundation of a course bearing the same name, and this was the core of EPA's technology transfer effort. The three-day course was general enough to educate Federal, State, and local officials, yet was specific enough to provide meaningful training to individuals wanting to enter radon measurement and mitigation businesses [1].

However, the ability of this course to meet the needs of many audiences was also its primary limitation. Different students had different informational needs. This realization led to the development of more specialized training programs: a house evaluation program, which gave students a basic, hands-on instructional experience in radon diagnostics and mitigation; and instructor training to ensure that as the radon health issue gained prominence nationally, knowledgeable trainers would be available to meet the rising demand for proficient service providers.

With the passage of the Indoor Radon Abatement Act in 1988, the U.S. Congress gave EPA the authority to establish Regional Radon Training Centers and a Radon Contractor Proficiency Program. Congressional intent was to have EPA ensure nationwide capability to provide information and training to Federal and State officials and the private sector on all radon issues. The Radon Contractor Proficiency Program came about as a way to evaluate the competence of radon contractors and provide this information to the public.

Early in 1989, EPA solicited applications through its ten Regional Offices from colleges and universities that were interested in hosting a Regional Radon Training Center. EPA developed guidance to assist institutions in understanding how the Agency envisioned the work of a Regional Radon Training Center. Three Centers were created in fall, 1989; a fourth was added in fall, 1990. The Regional Radon Training Center Network was established to provide specialized training to a wide variety of organizations involved with the radon problem, and specifically to aid in the development of a workforce capable of providing quality radon measurement and mitigation services to the public [2].

The core activity of the Regional Radon Training Center Network has been to support the Radon Contractor Proficiency Program. The program consists of several elements that collectively help to ensure the proficiency of participating radon mitigation and measurement contractors: (1) initial training course deliveries; (2) mitigation guidelines; (3) national proficiency examinations; (4) continuing education for the purposes of maintaining proficiency; and (5) re-examination every two years to maintain EPA listing (this provision changed in 1997 to permit accredited continuing education courses to suffice for continued listing without need for re-examination).

EPA has, until this year, provided States, Regions and the public with proficiency listings of mitigation and measurement contractors who have met the requirements of the program. The Agency distributed the report in both print and electronic database formats to all fifty States and EPA's ten Regional Offices. The Agency also provided copies of the report to environmental health organizations and popular trade publications. In this manner, officials responsible for environmental health issues, and the public, were assured that the service providers they hired to take radon measurements or perform radon mitigations were qualified to do the work in conformance with national guidelines.

This is an example of how the Federal Government synthesized technical information on an important emerging indoor air science topic, developed a series of training courses for both professionals and general audiences, established and supported institutions designed specifically to disseminate that information to those target audiences, and provided examinations so that qualified service providers could be listed by EPA.

3. Case Study: EPA's Indoor Air Quality Tools for Schools Training Modules

Elementary and secondary education is, according to the U.S. Government Accounting Office, America's largest public enterprise [3]. There are more than 80,000 public schools in about 15,000 school districts in the U. S.; there are additional 25,000 private schools. Over 42 million students attend public schools, and some 5 million more are educated privately. In total, 2.3 million teachers, 126,000 administrators, and 600,000 staff are employed to educate America's children from kindergarten through twelfth grade. This means that nearly one in five Americans occupies a school building each school day.

Indoor air quality has been a concern of EPA's since the early 1980's, when studies conducted by Agency and other researchers on people's total exposure to airborne pollutants demonstrated that the air inside buildings was often more polluted than the air outside. Mounting evidence suggests that school buildings may be among the poorest of building types with respect to indoor air quality. Based on a nationwide survey of schools conducted by the U.S. Government Accounting Office (GAO) in February, 1995, GAO estimated that 28,100 schools, housing some 15.5 million students, have less that adequate heating, ventilation, and air conditioning systems. As they reported in The Condition of America's Schools, 21,100 schools (with 11.5 million students) said they suffered from unsatisfactory ventilation, and 15,000 schools (serving about 8.4 million students) characterized their indoor air quality as unsatisfactory. Some school officials suggested that a major factor in the declining physical condition of the nation's schools has been decisions by school districts to defer vital maintenance and repair expenditures from year to year due to lack of funds. [3]

Identifying potential indoor air quality (IAQ) problems inside school buildings before students and staff suffer illness or discomfort, taking routine no-cost or low-cost steps to prevent problems from arising, and knowing how to fix them quickly if and when they do occur, is the subject of an action kit called *IAQ Tools for Schools*, published by EPA in 1995. The schools's kit is designed with the problems of schools in mind, i.e., the types of indoor air quality problems that are most often found in schools, and the kinds of prevention and problem-solving approaches that keep the typical school's limited budget in mind. *IAQ Tools for Schools* provides background information on what indoor air quality is all about and why it can be a problem, especially for schools. It contains easy-to-use checklists on topics such as ventilation, building maintenance, waste management, and renovations and repairs. It includes sample memoranda and policies, a model IAQ management plan, and a unique IAQ Problem-Solving Wheel. The kit also contains two fifteen-minute videos. The first is ATaking Action,@ a filmed case study of the successful adoption of the kit by an elementary school in New England. The second, A Ventilation Basic, shows how to walk through a typical school ventilation system, with special emphasis on proper operation and maintenance procedures. Each component is designed to be used by existing school staff. The kit also includes appendices on important topics such as mold and moisture control; basic indoor air measurement equipment; applicable codes and regulations; and, if, when, and how to hire outside assistance. All of the components are packaged in a manner that makes it easy to remove and replace reference materials, duplicate and disseminate checklists for staff, and add and update important information about a particular school's building systems or layout [4].

The kit was published with the support of six nationally well-known school and health organizations: the American Lung Association, the American Federation of Teachers, the National Education Association, the Association of School Business Officials, the Council for American Private Education, and the National Parent Teachers Association (PTA). Through these sponsor organizations, active outreach on the part of EPA's Regional Offices, and agreements with many other national environmental health organizations, interest in the *IAQ Tools for Schools* kit has steadily grown, and an increasing number of schools, school districts, and in some cases, whole States are considering adopting the practices outlined in this guide.

In response to increasing interest, in 1997 EPA initiated the development of two training modules related to *IAQ Tools for Schools*. One is designed to promote the importance of a good indoor environment in schools and the availability of a program to institute good practices. The intended audience for this module is quite broad, and it can be used with a variety of school communities ranging from parents, to teachers, to school administrators, to local and State government officials, indoor air quality consultants and service providers. The second module is designed specifically for the IAQ coordinator, the person (or team) designated in each school to administer the implementation plan contained in the kit. This module provides more detail on how to perform indoor air quality management tasks, e.g., distributing, collecting, and analyzing teacher checklists and building ventilation checklists. Each of the modules contains presentation and supporting materials for use by anyone interested in promoting adoption of the kit.

EPA is distributing the modules in several different ways: the primary mechanism is via CD-ROM; secondarily, through hard copy printed version; and in all probability, availability on the worldwide web. These materials are distributed free of charge to schools, local and State officials, and national environmental health and schools-based advocacy organizations.

In this example, the Federal Government designed indoor air quality guidance specifically for schools, developed course modules for two specific target audiences, and is making the course materials available to anyone who would like to promote adoption.

4. Case Study: EPA's Building Air Quality Training Courses

In 1991, EPA published a landmark guidebook on how to operate commercial buildings to prevent and solve indoor air quality problems. *Building Air Quality: A Guide for Building Owners and Facility Managers* was published jointly with the U.S. National Institute for Occupational Safety and Health. It contains practical advice for the people who run buildings about common indoor air quality problems and the key factors that cause them, routine measurements that can aid in managing for good IAQ, and easy-to-understand descriptions, checklists and forms, diagrams, and photographs of typical building equipment and problem situations [5].

The Guide was developed by EPA with the assistance of numerous outside experts to determine where consensus lay about best practices for building air quality management. Prior to its publication, the guidance was reviewed by a wide range of building community stakeholders. Among the key stakeholders was the Building Owners and Managers Association International (BOMA), a trade organization representing over 7,000 members who include office building owners, managers, developers, facility managers, and leasing agents. Its one hundred local affiliates collectively manage five billion square feet of office space in North America. BOMA supported EPA's development of the Guide and publicly endorsed it upon its publication. The organization wanted to encourage its members to adopt the practices in the guide, and to do so, requested that EPA fund the development of a training seminar which the trade association could offer to its local affiliates across the United States.

EPA collaborated with building science and adult learning experts to devise an introductory seminar of several hours duration, and BOMA worked with its local affiliates to promote the seminar to its membership. Over the course of about eighteen months, sixty local affiliates, including those in all the major metropolitan areas in the

U.S. offered the seminar which BOMA titled, Improving the Indoor Air Condition. It was a highly successful enterprise; local members paid a modest fee to attend the seminar and received a copy of the guide. Any proceeds that surpassed BOMA's costs for the instructors and space rental were used to offset the costs of additional seminar offerings. In this way, EPA's purposes of reaching a key target audience, building owners and managers, were met by working cooperatively with a trade association to deliver the message of improved indoor air quality management to its members.

Another highly successful training outgrowth from the *Building Air Quality Guide* came about as a result of interest on the part of the International Union of Operating Engineers (the Union), a labor union that represents among its 370,000 members, more than 100,000 stationary engineers. These engineers are responsible for the operation, maintenance, renovation, and repair of building systems. They are employed in schools, hospitals, hotels, industrial and manufacturing plants, and office and commercial buildings. In operating and repairing these facilities, stationary engineers perform work on boilers and steam systems; heating, ventilating, and air conditioning systems; building automation systems; pumps, piping and compressed gas systems; refrigeration and electrical systems; and numerous other physical plant functions. The operation and maintenance of these systems in large part constitutes the multiple influences which contribute to indoor air quality.

The Union proposed that EPA provide funding to support its development of an instructor manual and student workbook based on the principles of sound indoor air quality management contained in the *Building Air Quality Guide*. They assembled a team to guide development of the course, including indoor air experts, a curriculum specialist, and several member stationary engineers. Upon approval from EPA as to content and approach and a successful pilot test of the materials, the course was intended to become an integral part of the training program offered by the Union to its stationary engineers. Over the course of several years of financial support from EPA, the Union completed development of a seventy-five hour training course with both classroom lecture and hands-on components, pilot-tested it with key Union local chapters, trained thirty instructors who could present the course, developed a brochure and poster to market the course to both their Union members and the members' employers, established a lending library of related course videos, contracted for the development of a computerized IAQ presentation, and designed and implemented an automated testing/scoring system. To date, more than six hundred Union members have taken the course; these engineers are responsible for more than one billion square feet of commercial building space. The Union has fully integrated this course into its curriculum for stationary engineers, and is surveying members who have taken the course to determine the extent to which their on-the-job performance has changed as a result of this new knowledge. They have also begun to offer the course at local community colleges so that non-members interested in pursuing career opportunities in the field of building operation and management can have benefit of this knowledge as well.

These last examples of Federal Government roles in training in the indoor air sciences demonstrate how the Government can become involved in developing course materials and joining forces with an interested trade association to present them to their members, and in working with a labor union to allow them to develop course materials based on Government guidance and disseminate them through their own apprenticeship and training program.

5. References

1. Harrison, J., Salmon, G.L., Hoornbeek, J., Gillette, L., Price, D., and Fisher, G., (1990) Evolution of EPA's National Radon Mitigation Training and Quality Assurance Programs.

2. Salmon, G.L., MacKinney, J., Hoornbeek, J., Harrison, J., (1991) EPA's National Radon Contractor Proficiency Program.

3. United States General Accounting Office, Report No. GAO/HEHS-95-61 (1995) School Facilities: Condition of America's Schools.

4. United States Environmental Protection Agency, Fact Sheet No. EPA-402-F-96-004 (1996) Indoor Air Quality Basics for Schools.

5. United States Environmental Protection Agency, Publication No. EPA/400/1-91/033 (1991) Building Air Quality: A Guide for Building Owners and Facility Managers.

EDUCATION AND TRAINING IN THE FIELD OF INDOOR AIR SCIENCES IN BULGARIA

MARIA TCHOUTCHKOVA
National Center of Hygiene, Medical Ecology and Nutrition
15, Dimitar Nestorov Street, Sofia 1431, Bulgaria

Abstract

The problems of indoor air sciences and policy making are considered as an important part of the environmental health strategy. In this field of activity Bulgaria follows Health for All principles of the World Health Organization and the scope and purposes of the Environmental Health Action Plan for Europe (EHAPE).
The objectives:

> *To provide education and training at all levels so as to create specialists and teams of environmental health professionals who will be responsible for implementing and managing specific programmes to improve environmental health..*

> (EHAPE para 129)

> *To improve social and physical living conditions in settlements, particularly for the disadvantaged, in order to prevent disease and accidents and enhance the quality of life.*

> *To ensure information on the type and level of indoor air pollutants, especially in urban area.*

> (EHAPE, para 245)

are presented by the basis for action, priorities and actions of the Bulgarian National Environmental Health Action Plan (NEHAP) - the fundamental strategic document. It reflects the real situation, the multisectoral and interdisciplinary character of the problem "indoor air quality", the needs and possibilities in achieving the indoor environment, responding to its health equivalent.

Key Words: environmental health, indoor environment, education and training, multisectoral cooperation, living environment

The problems of indoor air sciences and policy-making are considered as an important part of the environmental health preventing strategy. In this field of activity Bulgaria

N. Boschi (ed.), Education and Training in Indoor Air Sciences, 191–196.
© *1999 Kluwer Academic Publishers. Printed in the Netherlands.*

follows Health for All (HFA) principles of the World Health Organization (WHO) and the scope and purposes of the Environmental Health Action Plan for Europe.

The emphasis on health promotion within HFA created the need to clearer definition of his strategies. The Ottawa Charter for Health Promotion, adopted in 1986, provided the strategic framework that was needed. It broadens the definition of health promotion, defining it as the "process of enabling people to take control over, and to improve their health". Enabling, mediating and advocating are the forms used to describe what health promotion does. Five elements make up the strategic framework provided by the Charter: promoting healthy public policy, creating supportive environments, strengthening community participation, improving personal skills and reorienting health services.

The principles of HFA, the strategic guidance of Ottawa Charter and the reorganization of our health services provide the framework for the indoor air science, education and training. Very helpful for our practice is the guidance document on "Strategic approaches to indoor air quality policy-making" (WHO European Centre for Environment and Health - Bilthoven Division, 1997) as a review of various dimensions of indoor air quality issues, addressing different indoor environments, pollutant sources, pollutants, exposure pathways and scenarios, health effects, prevention/mitigation options and confounding factors. This document provides also information to enable countries to choose among different indoor air quality policy options which will be the most effective, considering country-specific needs and resources. Finally, the document addresses the methods for performing risk comparisons, risk communication and quantitative measures to evaluate the effectiveness of indoor air quality policies on improved health and comfort.

The Second European Conference on Environment and Health, held in Helsinki, Finland, in June 1994, set the priority to the development of national environmental health action plans. The Conference endorsed the basic document for achieving this objective - the Environmental Health Action Plan for Europe (EHAPE).

It was decided to start with a pilot project, which includes six Member States - Bulgaria, the United Kingdom, Italy, Latvia, Hungary and Uzbekistan - as priority international action in support to the efforts of the European Member States, to develop their national plans. The selection of those six pilot countries at different stages of social, economic and political development, with different environmental health priorities and demographic problems and different geographic spread, reflects the necessity to gain a range of experiences that then will be shared with other countries. The aims of the project are:

- to assess the principles and strategies, outlined in the EHAPE;
- to check in practice the applicability and consistency of the directives and recommendations in the project;
- to serve as a constant source of knowledge and expertise for other countries;
- to give direction to environmental health activities in the spirit of the decisions of the Helsinki Conference.

Chapter 2.6 of the Bulgarian National Environmental Health Action Plan - "Professional Training and Education" presents the **objective:**

> *To provide education and training at all levels so as to create specialists and teams of environmental health professionals who will be responsible for implementing and managing specific programmes to improve environmental health..*

(EHAPE para 129)

Basis for action

Shortage of people suitably qualified for staff environmental health activities is a major impediment to improving environmental health management. Capacity building of environmental health professionals calls for a specialized training. Training should take account of geographical, cultural, economic and political characteristics and the multifaceted nature of the environmental health problems and requires a complex approach.

Environmental health professionals are medical doctors, environmental technicians, environmental/health scientists or technicians, construction engineers, chemists, physicists, architects, biologists, geographers, economists, lawyers, etc. Education of these professionals involves to different extent problems of environmental protection and mitigation of health risks due to environmental impact. Environmental health officers are educated to a degree level and trained in broad aspects of environment and health in terms of medical and technical knowledge, social policy, management and personal skills.

The concept of environmental management for the promotion of health is relatively new and underlines the need for appropriate education, either in the form of specialized degrees, or of supplementary professional education. Recognition of diseases occurrence associated with exposure to environmental factors calls for further education of medical personnel in environmental health. It is imperative to develop medical specialty on environmental health, in association with existing specialties as these of community hygiene, occupational health, sanitary engineering and sanitary chemistry.

Priority

- To introduce in schools, universities and other educational institutions, courses at various levels to educate and train environmental health and other relevant specialized personnel who can manage and facilitate the implementation of programmes in environmental health.

Actions

Introduce, in universities and other higher education institutions, programmes to educate and train specialists in environmental risk assessments and risk management.

Deadline: 1999

Actors: MoES, MoH, MoEW

Develop environmental health as a speciality with a suitable programme of continuing professional training.

Deadline: 1999

Actors: MoES, MoH

Increase coverage of environmental health in curricula for professional training in a wide range of subjects such as medicine, physics, chemistry, engineering, architecture, town planning, law, economics, social activities, etc., as well as for unemployed with appropriate qualification.

Deadline: 1999

Actors: MoES, MoEW, MoH,

MLSP

Provide continuous professional training for experts dealing with risk assessment and risk management in environmental health by organizing courses, workshops and other training activities.

Deadline: 1999 and constant

Actors: MoEH, MoH

Publish a code of regulatory acts in environmental health.

Deadline: 1999

Actors: MoEW, MoH

The issues that Bulgaria is facing now in the field of indoor air sciences and educational and training practice can be found in the Chapter 4.1. "Living Environment" of the Bulgarian NEHAP.

Objectives:

> *To improve social and physical living conditions in settlements, particularly for the disadvantaged, in order to prevent disease and accidents and enhance the quality of life.*

> *To ensure information on the type and level of indoor air pollutants, especially in urban area.*

(EHAPE, para 245)

No policy on settlement development has been elaborated in this country aimed at quality of life improvement. "The living conditions determine to a great extent the quality of life, which improvement is important for meeting the main needs of work, house, health service, education and recreation" (Vancouver Declaration on Settlements, 1976)

Many problems create the lack of clear-cut criteria and indices. The complex assessment of the environmental condition in the settlements is particularly difficult due to the complexity and multicomponent environment subject to study and control "man - physical environment - nature - urban environment - housing - indoor air quality - health risk assessment and management".

Housing is very important for the health and well being of the dwellers. Construction materials, type of heating, ventilation available and the household activity affect the indoor air quality. In the country there is neither a system for monitoring indoor air pollutants, nor studies on this problem.

At present the health perspective of the indoor environment is based on environmentally friendly management and development The inherent comprehensiveness of ecological knowledge, values and standards, of science and social practice highlight the road toward a favourable development of the health prevention in Bulgaria.

Priorities

- Creation, updating and harmonizing of the regulatory basis with the regulatory requirements of the EU.

- Developing transboundary cooperation between the countries in the Balkan region, integrating their efforts towards the improvement of the existing environmental situation and sustainable development of settlements.

- Solving problems related to the different interests and responsibilities of different agencies when decision-making on the management of the environment and on the health of the population

- Ensure and support living conditions which meet the health expectations of the population. Limit and reduce the morbidity and mortality from acute and chronic diseases connected with living conditions.

- Ensure sustainable development of areas and zones of high environmental risk - highly urbanized, with increased risk to health.

- Introducing systemic policy for protection and development of landscape and green wedges in settlements and around them.

- Improving the social conditions and the physical environment particularly for the poorer members of the society, decreasing the number of deprived and homeless persons.

- Reducing the negative impact on the environment caused by organization, processing and deposit of industrial and household wastes and discharge of waste waters.

- Rational use of natural and climatic resources for health promotion.

- Reducing diseases caused by indoor environmental factors (allergens, cancerogenes, air pollutants, physical factors).

- Creating well-educated staff with the aim of implementing the national policy for environmental health - decision making, selection of priorities, initiating and enforcing control measures.

- Improving public involvement in the development and implementation of the policy related to the environmental health (including health-related problems of indoor air quality).

The above-mentioned chapters of Bulgarian NEHAP are prepared by the author of this presentation - after serious consultations of all Departments and Institutions, regarding the multisectoral and interdisciplinary character of the problem "indoor air quality". Its influence upon environmental health, quality of life, and access of progress in the global issue - "sustainable development" requires much more efforts in creating an international and national strategy and plans for actions, including the education and training. Nevertheless, Bulgarian NEHAP reflects the real situation, the needs and present possibilities of our start in achieving the indoor environment, responding to its health equivalent.

Reference

Bulgarian National Environmental Health Action Plan, Ministry of Health, Ministry of Environment and waters, Sofia, 1998

EDUCATIONAL NEEDS IN EASTERN EUROPEAN COUNTRIES AND NEW INDEPENDENT STATES

LÁSZLO BÁNHIDI[1] AND VLADIMÍR BENCKO[2]

[1] Department of Building Services Engineering, Faculty of Mechanical Engineering Technical University of Budapest, Hungary
[2] Institute of Hygiene and Epidemiology, the First Faculty of Medicine Charles University of Prague, the Czech Republic

Key words: Indoor air quality, educational needs, indoor environment specifics

1. Introduction

Mainly two branches of studies, medical research and technical research deal with indoor air quality in closed spaces and its impact on human beings. The former chiefly concentrates on the impact on humans and diagnostic problems, while the latter examines the technical possibilities and conditions of ensuring optimal or permissible air quality from the humans point of view. The representatives of the two fields must work together in close co-operation: technical research have to meet the physiological and hygienic requirements of air quality and physicians have to be familiar with the technical possibilities and limitations of the specifications they have required and proposed.

Students of medicine, sciences and mechanical engineering must become familiar with all these requirements in higher education. The general situation of these topic, possibilities and necessary steps can be summarized as follows:

2. The necessary steps in higher education regarding the teaching of indoor air quality fundamentals

The problems of indoor environment, sick building syndrome and buildings related diseases especially in the context of allergies and hypersensitivity symptoms related to buildings are convenient subjects suitable for inclusion in subject hygiene (environmental medicine) when teaching students of medicine.

Besides of teaching of students of medicine only slightly modified courses on hygiene can be offered to students of the Faculty of Sciences on using the same teaching schedule and teaching textbook. When teaching the subject indoor environment related problems in this case can be just put less attention to health specific details.

It is recommendable to infiltrate to the technical university to extend contacts with architects and mechanical engineers, because this collaboration be very useful in

197

N. Boschi (ed.), Education and Training in Indoor Air Sciences, 197–200.

prevention of indoor environment problems. In teaching this kind of students it is useful demonstrate not only "frightening example: of indoor air history like legionnaires pneumonia, asbestos or radon carcinogenicity, but as well, at least some positive cases like potential beneficial influence of electroionizing microclimate and influence of different kinds of building materials in that context [1].

Now of all, we would like to present some of the typical problems of our countries, without trying to offer a complete overview.

3. Typical indoor air quality problems in the Central and Eastern European countries

When asked to comment on the differences in concepts and approaches to the problems of Indoor Air Quality we can frankly say our problems in Central and Eastern Europe, technically are as like as two peas of those in the West Europe and America [2].

Yet, the substantial differences can be seen in the practical applications. One difference is the persisting tradition of the habitual, natural ventilation. The situation is rather aggravated by our present unfavourable economic position, this being a major hamper for investments in air conditioning technology. In a current practice new or reconstructed buildings of banks and "multinationals" are exceptions of the rule.

We can be more specific now. When designing a new health care facility - nowadays rather its reshaping - we strictly insist on air conditioning in operation theatres, intensive care units and in a few more specialized health care units . Except for some big buildings which are unthinkable without air condition facilities the wards and patients' rooms are heated and ventilated by classical means -i.e. natural ventilation. For this reason the in-patient facilities are traditionally situated on the outskirts of the town, possibly in the green to ensure the air for ventilation as clean as possible to exclude potential risks for the patients' health. In fully air-conditioned hospital buildings there are often troubles with maintenance of air condition systems, mostly with their cleaning.

Further, we would like to present the following case. This year, in operation was put a new building in which our top centre of transplantation programmes, active well over 20 years, was moved. After a three months' run, in the freshly housed centre a serious of five deaths of pneumonia caused by Legionella pneumophila occurred in the kidney-transplantation ward. The source however, was not the air conditioning system. The hospital avails of a safe HVAC system using hot steam for moistening of the conditioned air. The origin of the problem was the distribution pipeline of hot water in which Legionella pneumophila had propagated to such an extent and reached the concentration sufficient for the necessary infectious dose in immunocompromised transplanted patients who used this water for washing, showers and personal hygiene. This situation is not some rare example and has been described repeatedly.

Another major difference between the West and e.g. the Czech Republic is the overheating of the houses. In the past , the price for energy in our country was government - subsidised and was much cheaper than today. The low-price heating, naturally, resulted in wasting and overheating of the apartments. The leaky insulation of windows caused massive heat loss so that our towns, under comparable conditions used to be by 1° to 3 ° of Celsius degree warmer than the towns of NATO countries. This was well visible from pictures taken from space in infrared spectrum.

The present day efforts to seal the leakage of windows and balconies, but before all application of leakproof plastic window frames are responsible for worsening of the natural ventilation in flats, and so there's no reason to be pleased. The quintessence of this world is simply dialectical. The real heating savings may be achieved only when we are able to regulate, individually, heat consumption in every room, and to shift the temperature throughout the day. Unless there are any adequate technological measures available for individual control in apartment houses (the said regulators do exist now in private family houses only) we will face the problems of only raised prices for heating. The current unfavourable situation is mostly due to the fact that many heating panels set in on maximum have the control valves seized, and out of order. The panels, or old types radiators, besides, have no measuring device for actual individual heat consumption.

After all, the permanently overheated flats with temperatures over 25 ° Celsius tend to produce potential health risks and makes the population much more touchy to cold. A higher room temperature is welcome by ladies who prefer cosy and comfortable indoor milieu. For physiological reasons it is desirable, mostly in bedrooms, to keep the temperature down, provided the inmates are no small children, babies or seniors, who often suffer from body termoregulation troubles. Another thing is the living room the temperature of which should be such, given the adequate clothing, as to suit everybody, even women.

Similarly to several other new technical problems, handbooks and standards in these education: related subjects e.g. ventilation, air conditioning and environment are taught as independent subjects or within indoor air quality in most of the countries at technical universities.

Requirements concerning the quantity of ventilated air usually fall into only two categories:
- in a smoking environment 30 m³ / capita fresh air
- in a non -smoking environment 20 m³/ capita fresh air
- Indoor air quality is separately defined by MAC numbers.

However, they are not taken into consideration in the planning of residential and office buildings (unless special materials are used. The olf-decipol values for emissions by humans, furniture, building materials etc., and the designing method are usually well-known, but they are only taken into consideration if requested by the investor as they are not included in the official standards for ventilation. Energy conservation measures also led to problems of indoor air quality. Windows and doors were designed that were almost air tight this did not ensure the necessary amount of fresh air. No artificial ventilation was designed at the same time.

4. Conclusions

Apart from economic rather than technical problems concerning large-scale applications of modern HVAC systems we have to cope with the basic questions of individual regulation (in our case mostly non regulation) or excess heating of the buildings.

With the teaching plans for the pregraduate and postgraduate students we have, at least, scored a relative success at the Faculties of Medicine and Sciences at Charles University and Faculty of Architecture at the Czech Technical University in Prague [1] and Hungary, Slovakia, Poland, Romania can demonstrate yet existing good examples of educational paradigmas in indoor environment related problems [3-6] Representatives from Bulgaria and Romania have presented structuralized proposals to start real work on this field. Representative of Poland demonstrated scheme of all building materials systematic testing from the aspects of potential indoor air problems including health risks and comfort conflicts rising from their use. In Hungary, Czech Republic and Slovakia are introduced practically the same testing schemes like in Poland. Another positive move was our share in the postgraduate training of hygienists, physicians engaged in public health and clinicians like alergologists [1]. Though our achievements in the training sphere should not be overrated they nevertheless appear promising for the future.

5. References

[1] Bencko,V., Holcátová,I (1998.)
Indoor Air Fundamentals and Graduate Education in the Czech Republic .Proc.CCMS /Budapest/NATO Workshop 1998

[2]Bencko,V.(1994) Health Risk of Indoor Air Pollutants A Central European Perspective.Indoor Environ.No. 3,213-223

[3] Bánhidi,L., (1998.) The teaching of indoor air quality at the Faculty of Mechanical Engeneering the Technical University of Budapest Proc. CCMS/Budapest, NATO Workshop 1998

[4] Ildikó Mocsy, (1998) " Educational Paradigms for Indoor Air Sciences in Romania".Proc.CCMS/Budapest,NATO Workshop

[5] Ingrid Senitkovs (1998) "Educational Paradiogms for Indoor Air Sciences in Slovakia" Proc.CCMS/Budapest,NATO Workshop

[6] Maria Tchoutchkova (1998)"Education and Training in the Field of Indoor Air Sciences in Bulgaria" Proc.CCMS/Budapest,NATO Workshop

EDUCATION IN INDOOR AIR SCIENCES IN POLAND

ANNA CHARKOWSKA, PH.D.
Warsaw University of Technology
Institute of Heating and Ventilation
Nowowiejska 20,00-653 Warsaw, Poland.

Currently there is no uniform educational paradigm in Poland, concerning indoor air sciences, adopted by higher level schools of similar profile of studies. It is a fact, however, that many subjects taught in various departments include the issues related to air quality, but generally the information provided does not constitute a coherent thematic block. Due to the lack of nation-wide curricula and consultations concerning teaching this subject, it is hard to ascertain which schools take up the issues of IAQ and in what measure they do it. The only regular event in Poland, related to IAQ issues, is the national conference entitled 'The Issues of Indoor Air Quality in Poland', organised every two years by the Institute of Heating and Ventilation of Warsaw University of Technology. In December 1997 the fourth conference in this series was held under the motto 'Healthy Home - Healthy Society'. Similarly to the former conferences, this one also aroused much interest which showed in a large number of participants representing various professions and in very lively discussions. In view of the participants' interest in the issues of indoor air quality, presented from various angles bearing upon their professional activities, it seems that the idea of formulating an uniform educational paradigm would be most welcome.

As a research worker in the Institute of Heating and Ventilation of Warsaw University of Technology, I can say with all certainty that in spite of the lack of such educational paradigm our students, who specialise in the design of ventilation and air-conditioning installations, get a large amount of information on IAQ, both as part of the teaching programme on a given subject and also as a separate important subject. These issues are taken up by the teachers who make their own programmes in 'Implementation of the Demanded Indoor Air Quality' and 'Protection of Air Against Pollutants'. Thus, although there is no nation-wide educational policy for professionals in the field of obtaining and preserving the demanded quality of indoor air, our students, who are the future designers of air-conditioning installations, get basic information on IAQ. It must be said, however, that this kind of education should be much more popularised.

Due to the specific character of educational system in Poland, teaching the subjects concerning the quality of indoor air may be discussed and implemented in some departments and on some levels. This teaching should be provided first of all for

N. Boschi (ed.), Education and Training in Indoor Air Sciences, 201–205.

professionals in building construction and related disciplines, that is in the Department of Environmental Engineering (ventilation, air conditioning, and heating installations), Department of Architecture, Department of Civil Engineering, and for the professionals in environmental medicine (biologists, chemists, physicians). I shall present in brief my suggestions concerning the teaching of the former group of professionals.

Technological education is provided in Poland on three levels, namely in:
- secondary schools of technology where the graduates receive the degree of technician;
- higher schools of technology (universities of technology) where the graduates receive the degree of master of engineering;
- graduate studies:
  - ♦ continuing courses for the graduates who will receive a diploma, but no scientific degree;
  - ♦ doctoral studies for the graduates who will receive the degree of doctor of philosophy.

Such a diversified system of technological education should be enriched by:
- introducing the subject which covers the IAQ issues into the curriculum (in the departments and the level of schooling where these issues have not been tackled hitherto);
- introducing interdisciplinary departments to provide teaching on IAQ for graduate and doctoral students.

The issues, which should be taken up, may be divided into four basic thematic blocks:
1. The demanded indoor air quality.
2. Effects of indoor air pollution on health.
3. Evaluation of indoor air quality.
4. Current legal regulations and proposals for new legal solutions.

Particular issues discussed within each thematic block may include:

Ad. 1. The demanded indoor air quality:

- sources and kinds of indoor air pollution, and methods applied to limit the pollution (for example, the choice of building materials characterised by limited or zero emission of pollutants into indoor air);
- indexes of indoor air quality;
- engineer's responsibility to provide good indoor air:
  - ⇒ decisions concerning the demanded indoor air quality should be taken at the stage of formulating the assumptions for the design of ventilation, air-conditioning, and heating installations for new or modernised buildings (they should include, among others, the permissible concentration of indoor air pollutants, the demanded volume of indoor air, treatment of air in conditioning units, supply air filtration);

⇒ actions taken during the operation of buildings and installations with the aim to preserve the assumed parameters and good indoor air quality;

⇒ commission of installations: check-up of the consistence with accepted technological and design solutions;

- role of the indoor-area users in providing good air quality;
- relations between air quality and productivity;
- method used for modelling the concentration of air pollutants in indoor areas and for predicting air quality.

Ad. 2. Effects of indoor air pollution on health:

- effects due to short-term and long-term exposure;
- illnesses due to air pollutants: prophylactics, symptoms, treatment;
- sick building syndrome and building related illnesses;
- methods of risk assessment.

Ad. 3. Evaluation of Indoor Air Quality:

- information gathered from the users' questionnaires;
- medical examination of the users' state of health;
- psychological examination to identify potential and independent sources of the users' dissatisfaction;
- identification and evaluation of the concentration of gas, dust, and micro-biological pollutants;
- examination of the indoor microclimate;
- efficiency measurements of ventilation installations;
- physical hazards measurements (noise, visual environment quality, ionising and non-ionising electromagnetic radiation, etc.).

The largest possible group of professionals (physicians, toxicologists, chemists, biologists, engineers, lawyers) should be consulted when developing the educational paradigm on IAQ, both at the conceptual stage and in the process of implementing the new programme of teaching.

The educational programme in the field of IAQ may take the form of a five-year interdisciplinary course of studies for the students particularly interested in IAQ issues. In the course of studies the students would attends classes held by various departments or schools; they would gain broad knowledge in various fields of science, apart from engineering. On completion the course of studies the students would submit a graduation thesis and receive master's degree.

Specialised graduate studies may be organised apart from the interdisciplinary course of studies. Two forms of such studies may be considered:
1. Obligatory studies for all graduates who are likely to encounter the problems related to IAQ in their professional career.

2. Optional studies for the graduates particularly interested in the subject (for example, those who had received their degree before the IAQ issues were introduced in the curriculum).

The studies may be run as:
- weekend trainings for a term or two terms;
- summer schools.

Apart from trainings held for the people who work on the design and exploitation of installations, open universities may be organised for small groups of people (20-30 persons), including the users who are dissatisfied with the quality of indoor air and who suffer from illnesses caused by bad indoor air. Such courses may enable them to see the causes of their health disorders and show them how to avoid such illnesses or prevent their growth.

In order to stress the importance of IAQ issues it may prove useful to devise an educational programme of cyclic events, such as:
- seminars;
- panel discussions;
- workshops;
- conferences.

Recapitulation

1. The first step to raise the social awareness of the issues related to indoor air quality is to provide thorough education for the professionals who create suitable conditions for the users of indoor areas.
2. Since there is no uniform educational paradigm on the issues related to IAQ, it is important, though not sufficient, to support any initiative which takes up these issues, even on a small scale (for example, thematic lectures as part of teaching the subjects already included in the curriculum).
3. In order to provide broad and well-organised education on IAQ issues, it is indispensable, however, to devise a nation-wide programme, adopted by every department and at every level of school education, which would be addressed to students likely to encounter problems related to IAQ in their professional careers.
4. Such education may be conducted on obligatory or optional basis and in various forms, ranging from separate interdisciplinary studies taking a number of years to single lectures held by invited professionals, which would be complementary to the current curriculum in various departments and at various levels of schooling.
5. When devising the educational paradigm on matters related to IAQ, it is necessary to take into account the people who occupy buildings. If need be, special trainings or lectures (for example, in the form of open university classes) should be offered to them in order to deepen their knowledge on the environment in which they live and on the prophylactics of illnesses caused by indoor air pollutants.

6. In order to propagate the knowledge of IAQ issues, it is vital to organise (not only in Warsaw) cyclic events, such as interdisciplinary seminars, conferences or panel discussions, which would provide an opportunity for the people to exchange opinions and discuss current problems.

7. Since the only event concerning the matters of indoor air quality and held regularly in Poland is the conference entitled 'The Issues of Indoor Air Quality in Poland', organised by the Institute of Heating and Ventilation of Warsaw University of Technology, it may be worthwhile to suggest that the organisers take up the idea of devising an educational programme on the IAQ issues as the theme of the next conference.

EDUCATION IN INDOOR AIR SCIENCES IN ROMANIA

I. MOCSY
Institute of Public Health Cluj
Str.Pasteur 6, 3400 Cluj-Napoca, Romania

1. Abstract

Indoor environment became an important subject after political and social change in Romania. Several laws were introduced in the debate concerning the public health. For this, the number of lesson in schools and universities were increased.

Much Non Governmental Organisations for environment were founded after 1990. These have a significant role in the education of the population. Unfortunately, in our country they still do not show the importance of the indoor air quality.

In Cluj, the Health Environmental Regional Organisation (HERO) organised trainings for schoolteachers and scholars in this field. It also made a short film about the "Radon in Schools", edited two booklets: "Radon in Homes" and " The Air Quality in Kindergartens". These were distributed in schools, kindergartens, and universities and in medical waiting rooms.

Our opinion is on the one hand to organise the training for architects, builders and teachers to intensify their education on indoor air and the other hand to increase their exposure to common international projects.

People spend 80 - 90 % of their time in closed spaces. In recent decades, homes, offices, work places, schools and public buildings have become primary habitats.. As the time spent indoors has increased, research has shown an incidence of respiratory tract diseases, allergies and other hypersensitivity disorders. Scientific interest has focused on the relationship between indoor environments and human health.

2. Introduction

International scientists have identified the following indoor air pollutant categories: Environmental Tobacco Smoke (ETS), radon, asbestos, organic, biological, inorganic, non-ionising radiation. These pollutants originate from a variety of sources, such as: cigarette smoking (a complex mixture of pollutants), combustion products (SO_2 gas, particles, organic), heating and cooking with gas (NO_2), vehicles traffic (CO),

207

commercial products: textiles, friction, cement, paper, plastic, insulation (asbestos) and paints, stains, adhesives, solvents, cleaners, pesticides, building materials and furnishings (volatile organic compounds VOCs and formaldehyde), ^{226}Ra content and physical properties of the soil under building and building materials (ionising radiation - radon), electric conductors (non-ionising radiation - electric and magnetic field), dust and biological contaminants (viruses, bacteria and moulds).

The major categories of indoor pollutants can be related to mortality, morbidity and reduced productivity. Lung cancer is the most serious health consequence that has been associated with inhalable particles (1, 2).

The National Statistical and Data Centre of the Health Department indicates a general increase of the death rate in Romania, especially caused by chronic lung diseases and lung cancer (3). Considering these data, the trends of deaths in our country over the last 20 years show following (Figure 1.):

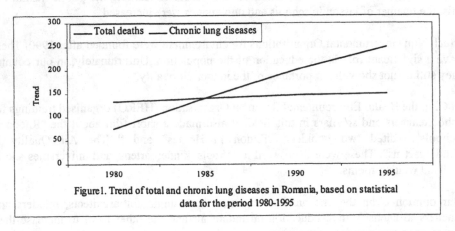

Figure1. Trend of total and chronic lung diseases in Romania, based on statistical data for the period 1980-1995

Therefore determining levels for individual indoor air pollutants has become increasingly important. After the political changes at the end of 1989, environmental problems were recognised as important issues which have direct impact on public health (4). Several laws regarding environment protection were newly adopted, e.g.
- "Quality in Construction" no.10/1995 5th paragraph: hygiene and environmental protection criteria should be satisfied in new constructed buildings;
- "Environmental Protection" no.137/1995 prescripts some requirements towards population protection.

3. Education on indoor air in Romania

Researchers from four Institutes of Public Health and 42 Inspectorates of Public Health follow different methods of measuring the pollutant exposure of humans in indoor workplaces, schools and homes. Several epidemiological studies were performed to determine the health effects caused by the exposure to polluted air.

The Indoor Air subject is present in various Universities from Bucharest, Cluj, Timisoara, Iasi, Craiova, Targu-Mures, Oradea and Arad, though in a very limited number of hours. Recently was introduced a new subject "Environmental Physics". Within this subject, the theme of the indoor air is treated as a significant topic. In spite of this, the students still do not have the possibility to get familiar with the practical application of the above theories. The TABLE 1. can show the departments and subjects that employed the "indoor air".

TABLE 1. Subjects on Indoor Air at Universities from Romania

University	Department	Subject	No. of Hour/year	Level of study
Medicine	General	Hygiene	5 - 8	5
	Dental	Hygiene	5 - 8	6
Technical	Construction	Building materials	70	1
		Science of heat	28	2
		Heating equipment	56	3
			70	4
		Ecology	14	3
		Ventilation &	126	4
		Climate	126	4
		Electrical equipment	28	4
		Environs. protection	70	5
	Architect	Equip. for building	56	3
Science	Physics	Equip. for building	2	4
		Atom physic	7	3
	Mathematics	Environmental physic	5	4
	Biology	Environmental physic	8-32	5
	Chemistry	Ecology	48	2-5
High schools	Nursing	Organic, An-organic (general) Hygiene	6	1

The number of Non-Governmental Organisations (3050) has significantly increased after 1990. Several of these have environmental issues as their main focus. These NGOs play an important role in the education of the population. Unfortunately, in our country, the importance of monitoring indoor air quality has been neglected (5). In Cluj, the Health Environmental Regional Organisation (HERO) developed different educational and training programs for schoolteachers and scholars in this field of indoor air quality. Documentation from the "NATO-Pilot Study on Indoor Air Quality" and "Indoor Air Quality - tools for schools" was important to these initiatives. HERO has produced a short film about the "Radon in Schools", and edited two booklets: "Radon in Homes" and " The Air Quality in Kindergartens". These information packets were distributed freely in schools, kindergartens, and universities and in medical waiting rooms.

The general population has no knowledge of the importance of indoor air quality. The risks of ignoring these issues include increased respiratory disease and death.

4. Assistance - proposal to extend existing educational activities concerning indoor air pollution

The goals established for educational activities concerning indoor air include characterising, understanding, and reducing the existing dangers to human health. I propose improving the educational activities in three directions.

4.1. EDUCATION FOR PROFESSIONAL DISCIPLINES

4.1.1. Many professional disciplines interweave because the indoor air quality issues have scientific, economical, and social implications. The elaboration of the strategies for prevention and mitigation of indoor air pollution adapted to the specific conditions in our country has been indispensable to organising the workshops and training in this field.

4.1.2. Convincing the authorities, that the exposure assessment for indoor air pollutants is one of the most important research activities. In several cases we have laws for indoor air pollutants but have inadequate equipment for determining them.

4.1.3. Internationally co-ordinated efforts have produced a series of scientific results. The researchers in this field have collected and shared the scientific results and experience during international conferences, congresses, symposiums and workshops. Unfortunately, only few Romanian researchers benefit from these. In the TABLE 2. we offer a brief picture of the Romanian presence at the most important indoor air meetings.

TABLE 2. Romanian participant at the International Conferences on Indoor Air

Title/Place of the Conferences	Date	No. of Romanian participants	Number of papers
INDOOR AIR			
Copenhagen	1978	-	-
Amerst	1981	-	-
Stockholm	1984	-	-
West Berlin	1987	-	1
Toronto	1990	1	1
Helsinki	1993	-	3
Nagoya	1996	2	6
Edinburgh	1999	?	?
HEALTHY BUILDINGS			
?	?	?	?
?	1994	4	12
Budapest	1995	2	2
Milan	1997	-	1 - ?
Washington			

I suggest increasing the participation of the environmental scientists at international meetings.

One of the possibilities to increase the awareness of the indoor air problem is to expend the co-operation of Romanian researchers with their colleagues from the more developed countries. This can be facilitated through conducting common projects (including students) and organising research exchanges for the harmonisation of the measurement and mitigation methods. The most experienced leaders of indoor environmental field lacked the necessary skills in proposal writing and successful project management. In the same time, the already implemented western solutions suggested to us for remediation have to be adopted to our conditions in order to be successful.

4.2. EDUCATION OF WHICH WHO CAN PREVENT THE INDOOR POLLUTION
(architects, building designers)

I feel it necessary to organise trainings and courses for architects and builders to increase their awareness of the importance of the indoor air pollution prevention. The impact of construction materials can be identified and taken into consideration within the concept of risk management.

212

4.3. EDUCATION OF THE GENERAL POPULATION

To inform the population about the indoor air risks is necessary to publish guidelines for indoor air quality and booklets presenting pollutants, health effects and methods for prevention.

NGOs have a significant role in the education of the people. Organising meetings will encourage contacts between the isolated NGO groups with indoor environment preoccupation, including raising public awareness about indoor environmental issues.

We believe that joining our forces with many professional disciplines from more developed countries, governmental institutions and non-governmental organisations, we will succeed in educating the different groups of the population on indoor air issues. We also hope that we can convince the authorities that the satisfactory indoor air is of vital importance for health, comfort, work and productivity and that it is an essential economic factor.

5. References

1. Maroni, M., Axelrad, R. and Tabunschikov, Y.A. (1997) Pilot Study on   Indoor Air Quality, NATO, Brussels
2. Sundall, J. and Kjellman, M. (1995) The air we breath indoors.
3. Anuar de Statistica Sanitara 1994 (1995) Ministerul Sanatatii, Bucharest, Romania.
4. Nicoara, S. and Mocsy, I (1997) Some aspects of Indoor Air Quality in Romania, *Indoor Built Environment* 6 224 – 231.
5. Catalogul Organizatiilor Neguvernamentale din Romania (1997) Fundataia pentru Dezvoltarea Societatii Civile, Bucharest.

CIB AND EDUCATION AND TRAINING IN INDOOR AIR SCIENCES

DR. PHILOMENA M. BLUYSSEN MBA[1] AND
DR. WIM BAKENS[2]

[1]: TNO Building and Construction Research
Department of Indoor Environment, Building Physics and Systems, P.O. Box
49, 2600 AA Delft, The Netherlands
[2]: CIB General Secretariat, P.O. Box 1837, 3000 BV Rotterdam, The
Netherlands

Abstract

CIB (Conseil International du Bâtiment or International Council for Buildings) is an
international membership organisation aimed at stimulation and facilitation of
international collaboration and information exchange in building and construction related
research and development. It comprises a network of 5000 building and construction
experts who improve their performance through international collaboration and
information exchange with their peers. In the CIB organisation there are about 60
Working Commission and Task Groups.

At the moment two Task Groups have plans with respect to training and education, one
of them in area of Indoor Air Sciences. This Task Group, TG28 "Dissemination of
Indoor Air Sciences" started this year and is a joint CIB-ISIAQ Task Group.

Introduction

CIB is the acronym of the abbreviated French name Conseil International du Bâtiment
(the English is: International Council for Buildings), CIB is an international membership
organisation. Members are organisations and also individual experts who are actively
involved in building and construction research and technology development. The main
part of the CIB membership consists of research Institutes, universities and industry and
practitioners.

With the support of the United Nations, CIB was established in 1953 as an association
whose objectives were to stimulate and facilitate international collaboration and
information exchange between governmental research institutes in the building and
construction sector. At that time an implicit objective was to help rebuild the European
infrastructure for building and construction research after the ravages of the Second
World War.

213

N. Boschi (ed.), Education and Training in Indoor Air Sciences, 213–219.
© 1999 Kluwer Academic Publishers. Printed in the Netherlands.

At the start 43 research institutes were members of CIB and by far majority these were European. And just as the programmes of these institutes, in the CIB programme likewise there was a strong emphasis on technical topics. Today, CIB comprises a network of over 5000 experts from about 500 organisations from research, industry and universities that cover all aspects of building and construction research and technology development.

CIBs aims and objectives

The Aim of CIB is to be an international association that stimulates and facilitates international collaboration and information exchange in building and construction related research and development.

The objectives of CIB are to be:
- a relevant and expert source for information concerning building and construction research and technology development world wide;
- a reliable and effective access point to the global research and technology development community;
- a forum for achieving a meaningful exchange between the building construction and the global research community.

Organisation of CIB

Figure 1 presents the organisation scheme of CIB.

The final responsibility for CIBs policy lies with the General Assembly which takes place every three years and in which representatives of all the CIB member organisations discuss and approve general policies and appoint the CIB President, other CIB Officers and the members of the CIB Board. The CIB Board then appoints the members of its two standing Committees: the Administrative Committee and the Programme Committee. The Administrative Committee deals principally with all financial and related affairs of CIB. The responsibility of the Programme Committee is for the CIB research and innovation related activities. It establishes the various CIB Commissions, it conducts an ongoing review of the performance of these Commissions and it terminates them if they have accomplished their task. Also, it officially appoints the Co-ordinators of these Commissions. The CIB organisation is facilitated by the CIB General Secretariat, which is located in the Netherlands (address: P.O. Box 1837, 3000 BV Rotterdam) and which operates under the supervision of the CIB Secretary General (Dr. Wim Bakens).

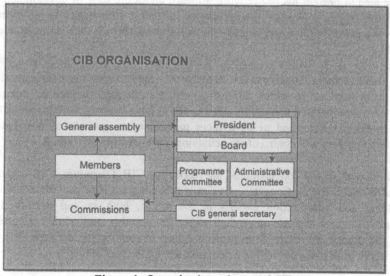

Figure 1: Organisation scheme of CIB

Activities of CIB

COMMISIONS AND TASK GROUPS

A CIB commission is a world-wide network of experts in a defined scientific area drawn from all over the world, who meet regularly and exchange information on a voluntary basis. The scope, objectives and work programme of each commission are defined by its members and officially approved by the CIB Programme committee. Most commissions have one co-ordinator, who is appointed by its members and by the CIB Programme Committee. Some commissions have not one but two joint co-ordinators and some have a secretary also. Some commissions have established working groups to focus on specific parts of the work programme and some have established their own projects in which commission members or their institutes collaborate in a more official way.

Although there are a few Commissions which only act as a platform for an informal exchange of information between their members, most also regularly produce publications.

In the CIB organisation there are about 60 working commissions and task groups. The complete list can be found at the CIB home page. Collectively these 60 commissions can in principle cover all aspects of building and construction research and technology development.

CONFERENCES AND SYMPOSIA

Most CIB Commissions organise an International Symposium or Conference every three or four years. The underlying aims are: to allow everyone working in the specific scientific area to present his or her work, to learn about the state of developments in all parts of the world, to update themselves on different approaches to problems they themselves are working on, and to make the initial contacts with their peers from all over the world.

A clear trend in the CIB community is to have meetings with the characteristics of a Workshop or Seminar in which all participants are asked to prepare an input that is focused on a well defined theme. The idea is not to have passive participants just sitting in and listening but instead to have them participate as actively as possible in the expectation that they will contribute to achieving a pre-defined final result.

Every third year, at the conclusion of the respective CIB Triennium, CIB organises its Triennial World Building Congress. An aim if this congress is to provide a platform for all building and construction experts in the world, both in and outside the CIB community, to meet and exchange their expertise. Another aim of this congress is to present and discuss all results from projects set up in conjunction with the CIB priority themes which took place during the Triennium.

PUBLICATIONS

Many of the publications form the factual basis for developing international standards or in themselves are used as such. Others are international state-of-the-art reports.

Three main types of published output originate from the CIB Commissions:
• scientific or technical analyses;
• international state-of-the-art reports;
• proceedings of workshops, symposia and conferences.

A Commission's project is often a scientific or technical analysis executed by a selected group of Commission members from different countries frequently with different scientific disciplines. A scientific analysis is of significance for the wider international scientific community and may be utilised for example in educational programmes in universities. The results of a technical analysis are intended for application by the building industry and practitioners.

A special type of Commission project is the international state-of-the-art analysis. The composition of its membership ensures that a CIB Commission is the most efficient vehicle for making such an analysis. The resulting publication is a report that describes state-of-the-art developments world wide in a way that is attractive and beneficial to researchers and educators as well as to innovators in the industry and practitioners and to governments.

Similar projects and publications of CIB Commissions are: the international status reports that analyse and compare developments per country and the international collections, and analyses of best practices.

Many CIB Commissions have an ongoing relationship with an international standardisation Commission, for example an ISO Commission. Through such a relationship a scientific or technical analysis or an international state-of-the-art analysis produced by a CIB Commission will often prove to be an effective tool within the context of preparing an international standard.

The most common CIB publications, however, are still the proceedings of CIB symposia and conferences which bring together refereed papers covering a defined area of building research and technology development by authors from all over the world.

CIB HOME PAGE

The CIB home page (address:www.cibworld.nl) is maintained by the CIB secretariat in Rotterdam, The Netherlands. This was started in 1996 and since them it has been developed continuously.

CIB and education and training

Recently, at the last CIB General Assembly (June 1998), one of the members pointed out that training and education is a growing field for all members. They used to be able to do only research for the government giving finances, but now it becomes more and more important to do what the industry asks and the industry asks more than research alone.

There is a shift from the industrial economy to a knowledge economy because of an increase in availability and amount of information. In the industrial economy the focus was to combine capital, equipment and workers (internal oriented). In the knowledge economy it is important to combine people with knowledge and technology, to create value for the customer and the market, and to increase the value of the company (external oriented). In the knowledge economy it is important to react fast on unexpected and unusual questions together with taking into account the different interests of the increasing parties. This asks for "value-based knowledge management" or an estimation of the value of the available information and something has to be done with it. One has to add value. Because of that there is a need for focussed transfer of knowledge and effective use of knowledge.

A few technical developments, which are happening parallel, are:
- The shift of use of IT in the learning process (computer based teaching) to communication technology. For example the use of internet;
- The development of other learning forms: a learning process that correspond maximal with the individual. An example of another learning form is the *virtual class* in which students learn on-line and interactive;
- As a consequence also, a need of knowledge management is arising with knowledge-based systems and databases to manage knowledge.

CIB has changed its name in June 1998 into International Council for Research, Studies and Documentation in International Council for Research and Innovation in Building and Construction. This implies a movement to the implementation aspects of research in general. In this context more attention for training and education is logical.

Also, it has been discussed because two commissions (TG26 en TG28) are working in this area, so it is time that CIB starts to formulate general starting points.

Task group TG 28 Dissemination of Indoor Air Sciences (joint CIB-ISIAQ Task Group)

One of the task groups that is focussed on training and education is Task Group 28 "Dissemination of Indoor Air Sciences". This task group is a joint CIB-ISIAQ Task group. ISIAQ (The International Society for Indoor Air Quality and Climate) is an international organisation which consists of members spread across many countries, disciplines and philosophies. For more information one can contact the secretariat of ISIAQ in Italy (isiaq@nemont.nemo.it).

The joint co-ordinators of Task Group 28 are: Mr. Gaute Flatheim (Flatheim AS, Norway) and Mrs. P.M. Bluyssen (TNO, The Netherlands). The members of this task group are listed in Annex A.

The original objectives and scope of Task Group 28 are:
- To focus on developing methods and activities/events for the transfer of the available scientific knowledge about Indoor Air Quality/Healthy Buildings to the building and construction community;
- To ensure that Hygiene, Health and the Environment are incorporated in the planning of new and in the rehabilitation of old buildings;
- To identify Target Groups to whom the knowledge should be addressed.

At the first meeting of TG 28 in Gaevle, at the CIB world congress, the group decided to do one or more of the following:
- Provide courses for international business executives;
- Development of CIB/ISIAQ guidelines to implement IAQ knowledge into the building process from beginning to end;
- Undertake a multi national pilot project with industrial partners.

The next meeting will be held in August 1999 in conjunction with the ISIAQ Congress INDOOR AIR'99 United Kingdom Edinburgh.

Annex A Members in:TG28 - Dissemination of Indoor Air Sciences (joint CIB-ISIAQ Task Group)

Mr. G. Flatheim M.Sc. Coordinator Norway

Mr. S.K. Brown Member Australia CSIRO - Commonwealth Scientific and Industrial Research Organization - Div. of Building, Construction & Engineering

Prof. M. Maroni Member Italy Università degli Studi di Milano - Istituto di Medicina del Lavoro

Dr. K. Motohashi Member Japan BRI - Building Research Institute - Ministry of Construction

Dr. J. Karbauskaite Member Lithuania Institute of Architecture & Construction

Dr.Ir. Ph.M. Bluyssen Member Netherlands TNO - Building and Construction Research

Mr. B. Bronsema Member Netherlands TVVL - Technical Association for Installations in Buildings

Prof. E. Trocka-Leszczynska Member Poland Wroclaw University of Technology - Department of Architecture - Institute of Architecture and Town Planning

Dr. J. Sowa Member Poland Warsaw University of Technology - Institute of Heating and Ventilation

Mr. J.L. Esteban Member Spain CSIC - Instituto de Ciencias de la Construcción Eduardo Torroja - Consejo Superior de Investigaciones Cientificas

Dr. C.G. Bornehag Member Sweden SP Swedish National Testing and Research Institute

Mr. C.H. Lu Member Taiwan - China Architecture & Building Research Institute - Ministry of Interior

Prof. G.J. Raw Member United Kingdom BRE - Building Research Establishment

Prof.Dr. B. Norton Member United Kingdom University of Ulster at Jordanstown - Built Environment Research and Technology Transfer

Dr. R. Edwards Member United Kingdom UMIST - University of Manchester Institute of Science & Technology

Dr. O. Barrat Partner Member ISIAQ Canada

Mr. E.M. Sterling Partner Member ISIAQ Canada

Dr. M.J. Paananen Partner Member ISIAQ Finland

Prof.Dr. O. Seppänen Partner Member ISIAQ Finland

Prof. C. Molina Partner Member ISIAQ France

Prof.Dr. K. Fiedler Partner Member ISIAQ Germany

Dr. P. Pluschke Partner Member ISIAQ Germany

Mr. L.G.W. Johnsen Partner Member ISIAQ Norway

Dr. S. Broms Partner Member ISIAQ Sweden

Ing. T. Follin Partner Member ISIAQ Sweden

Mr. B. Wessen Partner Member ISIAQ Sweden

Mr. M.S. Crandall Partner Member ISIAQ United States

Mr. W.M. Ewing CIH Partner Member ISIAQ United States

Dr. D.L. Hansen Partner Member ISIAQ United States

Mr. B.C. Krafthefer Partner Member ISIAQ United States

Dr. Ch.J. Weschler Partner Member ISIAQ United States

Prof.Dr. Ch. Sjöström CIB PC Liaison Sweden KTH - Royal Institute of Technology - Faculty of Architecture, Surveying and Civil Engineering

Dr.Ir. W.J.P. Bakens Liaison CIB General Secretariat Netherlands

ASSESSMENT OF INDOOR AIR QUALITY AND ITS IMPACT ON CHILDREN'S HEALTH

ANNA PÁLDY, A. PINTÉR, J. BÁCSKAI, P. RUDNAI, I. FARKAS* I. KATONA**, L. KŐRÖSI***, F. SZÍJJÁRTÓ**** AND K. PAPP

*"Jozsef Fodor" National Public Health Centre, National Institute of Environmental Health, ** City Public Health Institute, Százhalombatta, *** Outpatient Clinic, Százhalombatta, **** General practitioner, Szigetújfalu, ***** City Public Health Institute, Gödöllő

Abstract

An epidemiological study was carried out to evaluate indoor air quality in three towns with different level of ambient exposure and to assess the health impact of indoor air pollutants. The indoor air concentration of VOCs, formaldehyde and NO_2 was measured by passive monitors for 5 days in the homes of 183 schoolchildren in Hungary, in March 1995. A questionnaire was filled out concerning housing conditions, ETS exposure of children. Peak expiratory flow (PEF) was measured, the ratio of eosinophil cells was counted in nasal smear and the ratio of T and B lymphocyte subpopulation was determined.

Mean concentration of NO_2 was 42.8 $\mu g/m^3$ in homes with gas heating vs. 23.97 $\mu g/m^3$ with central heating (p<0.05) In the town with highest ambient VOC exposure ETS inversely correlated with PEF in children (r= -.24, p=.045). The indoor air concentration of NO_2 was associated with the presence of conjunctivitis r=.26, p=.033. Indoor benzene concentration was inversely associated with the proportion of B lymphocytes (r=-.30, p=0.048)

The ratio of eosinophil cells in the nasal smear was significantly associated with the diagnosed housedust mite allergy (multiple R= .33 adjusted R^2=.086 p=0.0039) Besides outdoor, indoor air quality has a considerable impact on children's health.

Introduction

In teaching indoor air sciences it is very important to underline the theoretical knowledge by practical results. Therefore it is unavoidable to carry out epidemiological studies in order to clarify the health impact of physical as well as biological indoor air pollutants.

221

N. Boschi (ed.), Education and Training in Indoor Air Sciences, 221–226.
© 1999 Kluwer Academic Publishers. Printed in the Netherlands.

222

The knowledge gained under country-specific conditions can serve as a basis for educational, as well as decision-making processes.

The aim of the presented study was:

1. Evaluation of the indoor air quality based on measurements of nitrogen dioxide, formaldehyde and aromatic hydrocarbon concentration.
2. Study of the health impact of the above mentioned pollutants on children living in this environment.

Methods

EXPOSURE ASSESSMENT

The indoor air concentration of NO_2, formaldehyde, and aromatic hydrocarbons was measured by passive samplers in the homes of 184 3rd and 4th grade schoolchildren in 3 towns with different levels of ambient air pollution in Hungary in March, 1995. (In the exposed town an oil refinery has been operating, producing a higher level of ambient aromatic hydrocarbon exposure). The ambient concentration of VOCs was also measured by the same method in each settlement. Passive samplers for sampling VOCs (SKC, USA) have been validated by comparing to active samplers and proved to be useful for 5-day long measurements. NO_2 and formaldehyde concentration was measured by photometric method, benzene, xylene and toluene concentration - by gas chromatography.

EPIDEMIOLOGICAL STUDIES

A cross-sectional study was carried out for assessing the health status in the study area and establishing relationship between air pollution and health outcome by the following methods: self-administered questionnaire containing 83 questions regarding housing: type of the house, floor, cover of wall, cooking, heating, presence of mould, smoking habits of the parents, history of allergic diseases of children. Peak expiratory flow was measured by a peak flow meter. The eosinophil cell ratio of nasal smear was counted. For the distinction of T and B lymphocytes in the peripheral blood smear the specific acid alpha-naphthylesterase (ANAE) (1) activity was demonstrated.

RESULTS

TABLE 1: Mean indoor concentrations of air pollutants

Pollutant $\mu g/m^3$	#	Mean	s.d.	Min	Max
Formaldehyde	181	12,75	8,52	2,7	46,9
NO$_2$	187	28,1	20,6	0,6	98,5
Benzene	157	9,82	11,44	1,1	77,6
Toluene	154	23,96	24,7	1,7	118,9
Xylene	154	20,39	21,31	0,1	123,2

TABLE 2: Mean indoor concentrations of air pollutants by location

town		Benzene	Xylene	Toluene	Formaldehyde	NO$_2$
I.	mean	8.5	20.2	24.9	14.1*	30.4*
	s.d.	7.73	20.9	23.8	7.93	20.4
	min-max	1.3-41.1	0.7-112	1.7-110	2.7-37.6	4.0-98.9
	median	6.15	17.1	18.2	11.9	24.2
II.	mean	8.28	17.4	22.6	9.3	19.6
	s.d.	10.5	12.3	21.5	6.93	13.3
	min-max	1.1-49.4	0.1-68.6	2.7-120	2.7-27.4	0.6-57.6
	median	4.2	15	17	6.9	18
III.	mean	15.64	37.3	24.2	13.7*	32.1*
	s.d.	27.2	45.7	28.5	9.5	22.4
	min-max	1.4-183	1.9-120	2.0-116	2.7-46.9	2.1-102.5
	median	7.3	17.5	12.8	11	25

* $p < 0.05$ ANOVA model

The ambient concentration of benzene was slightly higher in the town with oil refinery than in the other 2 locations, the difference, however, was not significant. The highest concentration in this location reached 20.6 $\mu g/m^3$, while in the other locations it was 6,2 $\mu g/m^3$ and 8.2 $\mu g/m^3$.

The types of the houses were: family houses (48.3 %), traditional house with flats (36.3 %) panel block (11.9 %). The major type of floor was parquette (36 %), floorcarpet was reported in 15 %. Most of the houses were built 10 to 20 years ago. Walls were covered with wallpaper in 58.7 %. In most of the houses there was central heating (in 57 %), and 18.4 % gas convector (with an exhaust pipe under the window). In 79.1 % of homes gas was used for cooking, and in 16.4 % - electricity. In 81.1 % of homes mould was not present.

The concentration of indoor air benzene had an association the settlement - it was the highest in town III. (multiple R= .349 adjusted R^2=.122 p=0.02) There was no association with the type of cooking, and heating. The association could be explained by an indoor activity, what was not recorded in the questionnaire.

Formaldehyde concentration was slightly associated with the type of cooking (multiple R= .14 adjusted R^2=.020 p=0.062). NO_2 had a significant association with heating (multiple R= .22 adjusted R^2=.045 p=0.0023) - mean concentration of NO_2 was 42.8 $\mu g/m^3$ in homes with gas heating vs. 23.97 $\mu g/m^3$ with central heating.

Peak expiratory flow was not associated with either of the indoor air pollutants. Opposite to general expectations there was no significant association between the number of smoked cigarettes at home and indoor benzene conentration. ETS exposure, on the other hand, inversely correlated with PEF in children living in the town with highest level of ambient air pollution (r= -.24, p=.045) In this location the indoor air concentration of NO_2 was associated with the presence of conjunctivitis r=.26, p=.033. Indoor benzene concentration was inversely associated with the proportion of B lymphocytes in this location (r=-.30, p=0.048)
Among biological indoor air pollutants housedust proved to be an important allergen. The ratio of eosinophil cells in the nasal smear was significantly associated with the diagnosed housedust mite allergy (multiple R= .33 adjusted R^2=.086 p=0.0039)

Discussion

Indoor air levels of toxicants are important in evoking adverse health effects, as for their concentration may exceed outdoor air concentrations. Since most individuals spend approximately 60-90 % of their time in indoor environments, this indoor air may constitute a significant source of exposure (2). There has been a growing recognition of the presence and diversity of volatile organic compounds in the indoor environment (3) which in high concentrations may cause exacerbation of bronchial asthma (4). For most volatile organic compounds the indoor concentrations exceeded the outdoor concentrations about twofold (5). In our study the indoor air benzene concentration in the location with oil refinery was about half of the outdoor concentration, toluene concentration was slightly higher indoor than outdoor. In town III, with the exception of 1 extremely high indoor air benzene concentration data, the median level was similar to that of location I.

The major sources of indoor air exposure to benzene by the general public are active and passive smoking, driving, pumping gasoline, and the use of certain commercial products (ie. marking pens, paints, glues, rubber products). In our field survey instructions were given regarding the replacement of passive monitors in order to avoid sampling biases. Data were not collected about using cars by the families, and about the place of garages. In an EPA (6) study benzene levels in the homes of smokers were 30-50 % higher than in the homes of non-smokers: the median daytime exposures were 21 μg/m^3 for smokers and 12 μg/m^3 for non-smokers. It was also found that in homes with garages or environmental tobacco smoke, mean indoor and personal benzene concentrations were two to five times higher than outdoor levels. We could only see a similar tendency in town II, in town I there was no difference between the two groups, in settlement III the mean benzene concentration was slightly higher in non-smokers' homes which data can be explained by other sources of benzene which were not defined in our study.

The effects of high levels of benzene are well known on hemopoesis (7), little is known about low levels of benzene exposure, but it may result in perturbation of the hemopoietic system (8). In our study the proportion of B lymphocytes showed an inverse correlation with indoor air benzene concentration in the town with oil refinery. The proportion of T helper lymphocytes was significantly higher and T suppressors lower than in the location III.

Housedust mites have been recognised as the major and in many areas as the major source of allergens in house dust. Mite allegy has been associated with perrennial rhinitis and asthma. It is very important therefore to take into consideratio the possible ways of reducing dust sources in the indoor environment at the level of planning and maintaining buildings. (9)
We could also detect the irritative effect of nitrogene dioxide. There are some studies indicating that rhinoconjunctivitis is much higher in residents living alongside roads with heavy traffic (10). It was also demonstrated that levels of night-time NO_2 increased in households with smokers and or gas fires, and that exposure of asthmatic children to these increased concentration of NO_2 correlates significantly with a decrase of their mean morning peak flow measurement the following day.(11)

Conclusion

Our results indicate the necessity that outdoor and indoor air pollution should be taken into consideration when studying the effect of air pollution on health. The study was sponsored by the EU Grant CIPD CT 930028 and the European Centre for Environment and Health, Bilthoven Division, The Netherlands.

References

1- Hulsse,C., Thielebeule, U., and Holzheidt, G. (1984) Zur Bestimmung von T-lymphocyten durch den Nachweiss der unspezifischen Esterase. Z.Gesampte Hyg.,. 30, 28-33

2. Fishbein, 1. (1992) Exposure from occupational versus other sources. Scand J Work Environ Health. 18. Suppl 1, 5-16.

3. World Health Organisation (WH0). Indoor air quality research: report of a WHO working group. Copenhagen: WHO, Regional Office for Europe, 1986. (Euro reports and studies; no 103).

4. Bent, S. and Zwiener, G. (1996) Solvent emission in a shcool building after using construction moisture protection substance. Gesundheitswesen. 58, 234-36.

5. Wallace, L.A. (1989). The total exposure assessment methodology (TEAM) study: an anlysis of exposures,,sources and risks associated with four volatile organic chemicals. J Am Coll Toxicol. 8, 883-95.

6. Cody, R.P., StrawdermanW.W. and Kipen, H.M. (1993) Hematologic effect of benzene. Job-specific trends during the first year of employment among a cohort of benzene exposed rubber workers. JOM 35, 776-82

7. Froom, P., Dyerassi, L., Cassel, A. and Aghai, E. (1994) Erythropoietin-independent colonies of red blood cells and leukocytosis in a worker exposed to low levels of benzene. Scand J Work Environ Health. 20, 306-8.

8. Platts-Mills, T.A.E. and Chapman, M, (1987) Dust mites: Immunology, allergic disease, and environmental control. J. Allergy Clin. Immunol. 80, 755-775.

9. Ishizaki, T., Koizumi K., Ikemori, R., Ishiyama, Y. Kushibiki E. (1987). Studies of prevalence of Japanese cedar pollinosis among residents in a densely cultivated area. Ann. Allergy, 58, 265-27.

10. Weeks, J. Oliver, J. and Carswell F. (1995). Respiratory effects of nitrogene dioxide exposure in asthmatic children. Eur.Respir. J. 8, 286S

PART VIII. A QUALITATIVE METHODOLOGY FOR DATA COLLECTION

FOCUS GROUP TECHNIQUES TO FACILITATE GROUP INTERACTION: FINDING A CORE CURRICULUM FOR INDOOR AIR SCIENCE

GABRIELLA M. BELLI
Virginia Polytechnic Institute & State University
7054 Haycock Road, Falls Church VA 22043-2311 USA

Abstract

Indoor air quality (IAQ) is of paramount concern to all human beings because we spend a large proportion of our lives indoors. Unfortunately, there is no unified indoor air science that addresses the problems associated with the IAQ of buildings from their conception through their lifecycle. In order to produce and keep healthy buildings, a number of different functional areas are involved, each with their own specialized knowledge base. Individuals involved in the production and maintenance of buildings do not necessarily have the tools to consider how their decisions may impact the environmental health and well being of the occupants. Likewise, the health professionals and exposure experts do not necessarily have the background to address building systems. An interdisciplinary approach to education and training is needed so that, at the very least, the different players are able to communicate effectively and together seek solutions for improving indoor environments. This paper outlines the use of focus group techniques in a workshop setting to facilitate the production of essential topics for a core curriculum for anyone involved in IAQ.

1. Introduction

Focus group interviewing is a qualitative data gathering technique that is widely used in social science research. Such diverse groups as advertisers, market researchers, educators, health professionals, program evaluators, and public policy makers use it. Its popularity is, in part, due to the fact that it is a rather flexible research tool that can provide in-depth information. They are particularly useful "for exploring the way particular groups of individuals think and talk about a phenomenon, for generating ideas, and for generating diagnostic information" (Stewart and Shamdsani, 1990, p.140). Focus groups may be used as a starting point in a larger research endeavor (e.g., Loneck and Way, 1997) or as a practical method of gaining feedback from a specific group. Participants are typically homogenous and often several similar groups are interviewed. For example, four focus groups of current and past students were used to evaluate and revise teacher education programs (Panyan, Hillman, & Liggett, 1997) and five groups of principals from one school system discussed their experiences with a school-based

N. Boschi (ed.), Education and Training in Indoor Air Sciences, 229–233.
© 1999 *Kluwer Academic Publishers. Printed in the Netherlands.*

management approach in order to provide both an informal evaluation and direction for a more formal one (Belli & van Lingen, 1993). Occasionally, dissimilar individuals are brought together in a focus group to facilitate constructive dialogue. For example, groups of nurse administrators, clinicians, educators, and students reacted to a changing job market and worked on adapting the curriculum to meet those changes (Morris, 1996).

In November 1998, a diverse group of 28 individuals came together at a NATO sponsored Advanced Research Workshop in Budapest. They represented 14 countries and a variety of professional backgrounds, including architects, engineers, medical doctors, physicists, chemists, psychologists, public health specialists, and policy makers. Yet they had one unifying factor, the arena of indoor air (IA) quality. One task for this workshop on *Education and Training in Indoor Air Sciences* was to generate a core curriculum that would detail what every IA scientist should know, regardless of his or her professional specialization. This task was an essential first step in bridging the communication gap between the people in a wide range of disciplines and professions. While a survey of IA professionals could have been used to generate a list of relevant topics, the combined effort of a group has more potential for producing a wider range of information. Additionally, it was deemed essential that a group diverse in both profession and nationality have an opportunity to interact and reach consensus on important aspects of defining a core curriculum. A modified use of focus group techniques was chosen by the workshop organizer to help provide a structure for the group interaction around this task. Specific details about focus group techniques, how they were used in this workshop, and how the group functioned are provided below.

2. Overview of Focus Groups

As normally conducted, a focus group is a formal, taped interview led by an impartial moderator. The group size normally ranges from 6 to 12. The moderator directs the session and guides comments, but the participants can move freely within certain guidelines. The process typically begins with participants reacting privately and in writing to an initial question. This independent idea generation is important so that the first few speakers don't influence the entire group. Each person, in turn, then presents his or her ideas. This is an equally important step because it allows for participants to build on each other's ideas and insures that a few people don't dominate the conversation. A discussion to clarify or elaborate on opinions raised may then take place. The ground rules specify that this should be a sharing of views in a nonjudgmental and noncritical way, with different opinions being important, and no arguments allowed. Once concluded, a transcript of the discussion is content analyzed by the researcher to extract common themes. The participant's role is to provide their original comments and then interact in a discussion and elaboration of the comments, but not to participate in the analysis. A number of books provide the theory behind and guidelines for using focus groups (e.g., Krueger, 1994; Morgan, 1993; Stewart & Shamdasani, 1990). Details about how this technique was applied in the IA workshop are related in the next section, along with a discussion of how the process differed from traditional focus groups.

3. Using Focus Group Techniques in a Workshop Setting

On the second day of the workshop, after having discussed the rationale behind developing a core curriculum and whom it would be for, the group split into two working groups of about a dozen people each. The first question they were to address was "What topics/concepts should be in a common core that all IA professionals should know?" Once formed, the two groups adhered to the basic ground rules of focus groups. They each wrote down what they considered to be the most important topics for a core curriculum in indoor air science. Each person then presented these in turn. The discussion that followed was nonjudgmental and built on each other's presentation. As expected, the moderators guided the proceedings and made sure that each person's views were heard.

However, the moderators in this case were active participants who not only directed the sessions, but who also helped clarify and organize input. Comments were recorded on flip charts by the moderator in one group and by a separate recorder in the other. In essence, the groups did their own initial content analysis as part of the group process. Both groups produced a list of about 60 topics and each one independently organized their list into a smaller set of nine themes or keywords. Then the two groups reunited to compare notes. Although the specific words differed to some extent, the consistency in meaning across the main categories was amazing and the participants were quick to agree about the few topics unique to one or the other list. The next step, after some discussion, was for a couple of participants to enter the categories and their sub-sets into a computer file for distribution to the group. The result was the mutually agreed upon topics necessary for a core curriculum. Subsequently, a few participants did the editorial work to merge overlapping subcategories and generally refine the lists. This product of the workshop is presented in the chapter dealing with the results of the first focus group sessions and the core curriculum.

Two more sessions using focus group techniques took place during the workshop. These dealt with perceived obstacles to implementing a common core curriculum for IA scientists and ways of overcoming those obstacles. After the groups provided their input and deliberated separately, they regrouped to view each other's results. The most salient points were then determined by a group vote, where each participant selected what they considered the top two priorities. Items were retained only if a number of individuals ranked them among the top two. Although not as consistent as with the topics, the two groups produced a number of similar obstacles and solutions.

4. Group Cohesiveness and Diversity

The whole group session leading up to the first set of focus group interactions was much more contentious than the focus group sessions. This discussion provided another difference from traditional focus groups because participants in the workshop interacted extensively to precisely define the task prior to splitting into two groups. One group even spent at least 15 minutes in a continued debate of defining the task before proceeding to the focus group format and identifying core topics. Based on a careful review of the audiotape transcripts of this whole group session and the focus groups, it was evident that different viewpoints prevailed. One viewpoint was that different content topics were needed for each specialization involved in IA. A second view, consistent with the goal for the workshop, was that a common core of topics was needed for all IA professionals. Additionally, some members of the group saw the target audience for a common core of knowledge and understanding about IA to include policy makers, teachers, children, and the public in general. The resolution was an acknowledgment that, while different audiences need to be educated, the immediate focus would be on creating a core curriculum for people with an academic degree within building design, construction, operation and maintenance, and investigation, as well as for occupational/environmental health and general practitioners.

Although consensus was reached about moving forward with the group work to develop a core curriculum, the different viewpoints did not disappear. One reason consensus was possible was recognition that this initial step would define the parameters of topics everyone needed to know at least at some level, however basic. There was agreement on the need for a common vocabulary and an interdisciplinary approach to IA education. How deeply each topic was developed or how much emphasis each would have relative to other topics would have to be situation specific. It would need to vary depending on discipline, on audience, and on duration and nature of a course.

It seemed that the participants needed to have this preliminary whole group discussion and to jointly arrive at a consensus about exactly for whom a common core curriculum would be developed at his time. Opinions were too diverse to initially agree on a specific target group. It was recognized that different approaches were needed for academic IA scientists, for working professionals with different educational requirements, and for the public at large who can be both victim and perpetrator of sick buildings. Added to this was the fact that different countries have different policies and educational systems, making it difficult to introduce new courses. All together, this made it quite difficult to find common ground. Yet, once all these aspects were on the table and it was understood that all were seen as relevant, agreement was easily reached. The first step was to determine and agree on a valid core of common knowledge for professionals involved in all aspects of IAQ. This was accomplished with the first set of focus groups. It was further agreed that results could then be modified and applied at any level and for any audience. But this task was left for the future.

5. Summary

Qualitative data gathering tools are not intended to produce generalizable results and it is generally understood that focus group results should not be generalized. However, a main purpose at this workshop was to produce a list of topics for a core curriculum that could be promoted at least to the larger population of indoor air scientists. A modified focus group approach worked well as a tool to facilitate this task. It provided a slightly more formal structure than the workshop's whole group interaction and gave every participant an opportunity to have input. The degree of consistency in the final lists of essential topics produced by the two groups working independently provides a sense of validity for the resulting topics. The relative ease with which the final product was agreed on is an indication that, at least at the core topical level, there is a remarkable amount of agreement across IA scientists from a number of fields, professions, and countries. What remains to be done is the development of these topics with proper emphasis for different specializations and with different formats and depth for diverse audiences, from graduate students to schoolchildren and from professionals to the general public.

References

1. Belli, G. and van Lingen, G. (1993) A view from the field after one year of school-based management. *ERS Spectrum,* Winter, 31-38.

2. Krueger, R.A. (1994) *Focus Groups: A Practical Guide for Applied Research.* Sage Publications, Newbury Park, CA.

3. Loneck, B. and Way, B.B. (1997) Using a focus group of clinicians to develop a research project on therapeutic process for clients with dual diagnoses. *Social Work,* **42**, 107-111.

4. Morgan, D.L. (1993) *Successful Focus Groups: Advancing the State of the Art.* Sage Publications, Newbury Park, CA

5. Morris, R.I. (1996) Preparing for the 21st Century: Planning with focus groups. *Nurse Educator,* **21**, 38-42.

6. Panyan, M.V., Hillman, S.A., & Liggett, A.M. (1997) The role of focus groups in evaluating and revising teacher education programs. *Teacher Education and Special Education,* **20**, 37-46.

7. Stewart, D.W. and Shamdasani, P.N. (1990) *Focus Groups: Theory and Practice.* Sage Publications, Newbury Park, CA.

PART IX. SUMMARY AND CONCLUSIONS

NEW DIRECTIONS FOR EDUCATION IN INDOOR AIR SCIENCES: AN INTERDISCIPLINARY AND INTERNATIONAL UNDERTAKING

N. BOSCHI

Virginia Polytechnic Institute and State University
7054 Haycock Road, Falls Church, VA 22043-2311 USA

Introduction

There are a number of actions that are taken to improve the environmental quality of living spaces; however, we still have unsatisfactory results. Education is an important component in the prevention and control of adverse health effects caused by indoor environmental exposures. Considering that many actions to improve the quality of indoor environments are dependent on those designing, operating and maintaining buildings, as well as on the occupants, the need for emphasis on education becomes evident. By providing appropriate education and training for professionals involved in the building and health sector and by raising public awareness, prevention of health effects associated with indoor environments becomes realistic. Moreover, it may offer a cost-effective way to reduce morbidity (e.g., pneumonia caused by Legionella pneumophila) and mortality (e.g., cancer caused by radon).

This chapter presents the results of this project focused on defining a framework for education in indoor air sciences (IAS). The results here summarized are organized in six points: (1) overview of existing programs, trends and new issues that are leading education in IAS; (2) audience addressed by the curriculum; (3) educational framework; (4) action plans for implementation of the core curriculum; (5) recommendations; (6) discussion and conclusions.

1. Existing programs and new directions

Contributors profile this scenario quite extensively in Parts I - VII of this report. Some of their points are summarized here. Today, there is not a specific "indoor air science". Scientists and professionals working within IAS come from a wide range of academic and technical fields, and they are educated and trained in indoor air within already established disciplines, such as mechanical engineering and occupational health.

A number of academic programs that aim to bridge the existing gap between health and building sciences are being offered at institutions of higher education in North-America, Australia, Europe, and Scandinavia. Further, a number of government and non government agencies offer or are in the process of offering educational programs to raise public awareness and reach the building and health professionals that are already practicing in this areas. Although in different regional contexts, lack of awareness of the

N. Boschi (ed.), Education and Training in Indoor Air Sciences, 237–247.

magnitude and character of the indoor air problem still causes resistance to education in IAS.

The programs presented aim to overcome the barriers related to the different approaches taken by the different professional practices involved in IAS. Also, they show that the major reasons for the lack of cross-disciplinary communication and understanding are the lack of an agreement on common goals, ethics and responsibilities and a common vocabulary.

New issues that education in IAS is facing were introduced. Environmental sustainability will change the approach to the way buildings are designed, built and operated. Traditional professions, such as engineers, are facing a change in their strictly technical roles. Occupants' assessment of indoor environment will assume a central role in building diagnostics. Transfer of knowledge and technology will enable global education and training based on clear understanding of the philosophy of teaching in IAS as well as an understanding of local needs.

Overall, a need for a new educational paradigm that recognizes "generalized knowledge" as important as "specific within-science knowledge" was identified. A need for holistic evaluation of cross-disciplinary research and scientific efforts was emphasized. Moreover, the need for multidisciplinary education was introduced as requiring harmonization at the international level while recognizing cultural and local diversity.

2. Defining the audience

This phase focused on creating a forum for the participants to interact and reach consensus on important aspects of defining a core curriculum. Participants discussed the need for developing a core curriculum for IAS, the rationale for developing a core curriculum, the audience to be addressed by the core curriculum as well as the criteria to state the curriculum. The different perspectives and issues raised by the participants constituted the stepping stone for the discussion in defining the workshop tasks.

The group reached consensus on the need for developing a core curriculum for indoor air sciences. The rationale for developing a core curriculum and the audience to be addressed were defined. The definition of the audience was obtained through extensive discussion. The core curriculum was stated as follows: "*The core curriculum applies to: all people with an academic degree within building design, construction, operation and maintenance, investigation, and all occupational/environmental health and general practitioners*".

After having addressed those receiving academic degrees, the matter of discussion was related to all those people who influence buildings who never go to universities. It was agreed that is a separate issue because the vehicle by which you reach that population varies tremendously from a university atmosphere, what is needed is to identify where those people get information that help them do their job and work through those organizations. It was recognized that it is difficult to identify a ready-made delivery mechanism. Probably all is drawn from the core curriculum and by starting there it will be possible to know which of these concepts would be optimal for a manager or a civil manager.

The group, considering the different disciplines and professions involved as well as the diversity in the level of knowledge needed to carry out their tasks related to indoor air, agreed on defining the curriculum as being stated in terms of topics.

3. A framework for education in indoor air sciences

The third phase of the workshop consisted of three subsequent focus group sessions. The questions discusssed in the three sessions were: *"What topics/concepts should be in a common core curriculum that all indoor air professionals should know?"; "What are the perceived obstacles to implementing a basic core curriculum on IAS?"; "What are the possible strategies to overcome the obstacles?"*.

THE CORE CURRICULUM

Results of the Focus Group Session 1. The main purpose of this session was to produce a list of topics for a core curriculum that could be applicable to the larger population of indoor air scientists. The participants, divided into two groups, identified a list of approximately 60 topics and each one independently organized their list into a set of topics or keywords. When the two lists were compared, in a common forum, the participants noted the consistency of the main categories. Agreement on the main topics was reached very quickly. The resulting common list was organized in the following *Primary Elements* (overlapping non-exclusive): people, buildings, sources, exposures and effects; and *Tools* (to be applied to the primary elements): scientific, regulatory, vocabulary and ethical principles.

Participants discussed and agreed that the specific emphasis of the topics was very important, however, it would need to vary depending on level of audience and program. Having defined that all people who will either design and construct buildings, make decisions about how they run or those who diagnose ailments that are building related, at some point of their academic education, should have knowledge of the core curriculum, the question that was discussed was about when that point comes. It was recognized as a difficult one to address because very dependent on the educational system within which that person operates. And that can vary from country to country. However, by establishing the content, people can work within their own system or within their own professions to say "well this is the appropriate point in this mechanical engineering program" when those concepts should be introduced. But that might be different in different countries.

A summary of the topics that identifies the core curriculum is shown in the tables that follow. The summary was developed from the extensive list of topics provided by the participants, although, the interpretation here presented reflects only the author's perspective. The first 2 tables contain the key words, and main related topics, organized in *Primary Elements* and *Tools*. The third table summarizes a set of interrelationships existing between the larger environment, buildings and occupants.

240

Table 1. The Primary Elements of the core curriculum: people, buildings, sources, exposures and effects and main related topics.

Key words	Related Topics
People/Building occupants	Physiology of main human targets, psycho-social factors in relation to environmental exposure; prediction of future problems by monitoring social and psychological trends; health economy; (other topics include: communication; overcrowding; endpoint effects, and measurements of the traditional variables related to exposure/uptake effects)
Buildings	Building siting; building sciences; building physics; building process, from design to operation and maintenance; housekeeping practices; response to identified problems (other topics include: ventilation; pollution pathways; controls; material selection; building technology; automation; management; economics and economic end points; potential risks from building and its equipment; solutions to known problems; building safety; building security; special requirements in special buildings; and indoor air process: factors leading to air deterioration or factors leading to air amelioration)
Sources	Physics of source emissions; product and material development; potential risks from building techniques (other topics included: new sources and new effects; asbestos and MMMF, and Tobacco smoke, humidity in building and thermal environment)
Exposures	Chemistry of indoor environment; toxicology of pollutants; physical environmental factors (other topics included: biological; particulate; aerosol physics and other exposure in general)
Effects	Human response (i.e., toxic, immune system and psycho-social); health effects (e.g., short term: irritants and long terms: allergens); harmful factors; potential risks from occupants and their activities; prevention and cure

Table 2. The Tools of the core curriculum: scientific, regulatory, vocabulary and ethics. The tools are seen as those means available to define, intervene and address the indoor air process.

Key words	Related Topics
Scientific:	Knowledge, tools and practice; building ecology; environmental health science; physiology and medicine; exposure assessment; risk management; engineering (other topics included: building diagnostics; monitoring; questionnaires; monitoring social and psychological trends management of indoor environment; influence of care and maintenance)
Regulatory	Laws; regulations; standards; guidelines; control and prevention approaches (other topics included: handling problem buildings; hazards and good practice in building specification and design; and Policy making)
Vocabulary:	Communications between and among disciplines and languages (e.g., terminology and abbreviations)
Ethics:	Ethical principles and common set of goals

Table 3. The interactions existing between the larger environment, buildings and occupants. The study of these interactions is leading toward environmental models that are more complex.

Key words	Topics
Interactions:	History of built environment; building ecology; building–environment; building–occupants; influence of design on comfort and health; impact of material selection on health and comfort; indoor air process: factors leading to air deterioration; factors leading to air amelioration; building physics; impact of material selection on health and comfort; and health economy.

The core curriculum was identified, although, the topics need to be further developed with proper enphasis on different specializations and with different formats and depth for diverse audiences.

PERCEIVED OBSTACLES TO IMPLEMENTING A BASIC CORE CURRICULUM ON INDOOR AIR SCIENCES

Results of the Focus Group Session 2. The main purpose of this session was to discuss and identify possible barriers to the application of the core curriculum. The opinions that came up in this session emphasized the existing differences within each country in terms of policy and educational systems. However, five main categories of obstacles were identified:

1. The players do not perceive IAS as relevant to the course being undertaken

    a.   Society does not see IAS as important

    b.   Political will does not exist, even if officially the Government has made a commitment (for example there may be no governmental authority, or no single governmental authority, which is responsible for IAS)

    c.   The NATO working group is just a minority, making trouble by pushing its ideas on everyone

    d.   There is no compelling reason to learn IAS if there are no external requirements relating to a particular discipline (e.g. codes, standards) or standards may actually conflict with good IAS teaching

    e.   IAS is not a required topic for the degree to be accredited (e.g. by professional body, Government, faculty)

    f.   Faculty members and/or students see IAS as irrelevant to their present or future profession: where is the career in IAS? There is no example of what someone with a qualification in IAS will do with his/her life

g. There is not a common goal between disciplines, therefore, it is difficult to agree on an approach to meeting goals. Related to this, there may be quite different professional incentives and career structures

2. Even if it is seen as relevant, key individuals are resistant

a. Individuals (e.g. lecturers, heads of department) feel threatened because they personally do not have sufficient knowledge of the field: Students will know more than they do

b. IAS threatens to dilute the "purity" of the discipline or scientific procedures: professional egos are damaged by putting "inferior" academic material in the syllabus

c. Co-operation between faculties/disciplines is needed, where there may currently be conflicts (e.g. over resources) or lack of trust

3. The will is there but the academic structure does not easily adapt to incorporate IAS

a. Courses have run in the same way for a long time and it will take effort to re-organize them

b. There is a lack of common technical language among professionals

c. Each discipline has its own teaching/learning style, jargon, etc.

d. The quality control that exists within disciplines does not automatically extend to the IAS component of the syllabus

4. The will is there and the structure can be changed but the resources are not sufficient

a. It will be difficult to find the time required for the extra teaching

b. Even if the time is available in theory, there are logistical problems scheduling that time for staff, students, and facilities

c. There is a lack of teaching tools and instructional materials - don't exist yet for each of the disciplines in the appropriate depth, style (e.g. mathematical vs. verbal vs. pictorial) and conceptual structure

d. Materials may exist in some languages/cultures but who will translate and transpose the material?

e. There is a lack of good teachers/professors of IAS: this may limit credibility even if some kind of course can be mounted

f. In some cases there may need to be a financial transfer between faculties: where will faculty A find the resources to pay faculty B if there is not an equal exchange? By cutting staff from faculty A?

g. Faculties may be geographically some distance from each other

h. If one of the solutions is to use distance learning to facilitate teaching, the required technology may not exist

5. Blockages in NATO working group

a. Our curriculum would be adopted if only we would write it and get a systematic approach to promoting it

POSSIBLE STRATEGIES TO OVERCOME THE OBSTACLES

Results of the Focus Group Session 3. Building on the results of *Focus Group Session 2*, four key obstacles to be addressed were selected: IAS is not a recognized discipline; if IAS were recognized as discipline how will quality control be exercised at the professional level; need for political commitment and lack of public awareness. These obstacles were associated with three key players or primary environments of operation: academic, professional, and policy settings and possible strategy were developed as follows:

Possible Strategies in academic environment

Obstacle: "We" are a borderline group; we don't belong to any one discipline - Indoor Air is on the outside.

Strategy:
- Create an independent discipline: Indoor Environment

- High jack an existing discipline or

- Merge with environmental sciences

- Hygiene is an existing discipline (in some places this not a problem because there is opportunity)

- Develop a common language; understand languages of others and serve as a bridge

- Define who "we" are

- Raise visibility and stature in Academia

- Make a list of available opportunities for doctoral students, post-doctoral students (e.g. environmental medicine specialists within occupational medicine)

- Examine local context and apply whatever of these strategies will work. Terminology and opportunities may vary

- Articulate indoor air concerns in a manner that is accessible/acceptable (both within academic setting) "Aggressively deliver polite messages"

Possible strategies in professional environments

Obstacle: How will quality control be exercised/professional code of conduct (ethics).

Strategy:
- Create a model code for the different professions

- Increase scientific rigor both in methods and in questions

- Make sure that excellence/research quality drives conferences on IA, not the numbers of attendees

- Create or entrust a Board to ensure quality

- Encourage publication within each profession's literature

- Get Indoor Air Journal increased recognition

Possible strategies in policy settings

Obstacle: Lack of Political commitment and lack of public awareness

Strategy:

- Remind politicians of statements already made and accepted by credible organizations

- Develop new/revised statements

- Enlist support of other bodies e.g. NATO (recommends to environmental Ministries), American Society of Heating, Refrigerating, and Air Conditioning Engineers (ASHRAE), World Health Organization (WHO), International Academy of Indoor Air Sciences (IAIAS), International Society of Indoor Air Quality and Climate (ISIAQ)

- Better publicize risks of indoor air and identify specific problems (e.g. formaldehyde, CO) as indoor air

- Target influential groups to lobby the politicians on IA importance. E.g.:
    - Non Governmental Organization
    - Health advocates
    - Patient groups
    - Consumer safety groups

4. Action Plans for the implementation of the core curriculum

A plenary session followed the focus group sessions. As group a number of action plans were identified. The most salient points were determined by a group vote, where each participant selected what they considered the top two priorities. Items were retained only if a number of individuals ranked them among the top two. Following are the action plans identified:

- Develop further the results of this NATO workshop into a series workshops, sponsored by NATO, on language, target audiences, specific subject matter, and obstacles

- Identify an action plan with responsible persons

- Create a framework/infrastructure that focuses on education rather than for science, perhaps using existing entity such as: International Academy of Indoor Air Sciences (IAIAS), International Society Indoor Air Quality and Climate (ISIAQ), NATO or United Nations

- Needs a leader, the institution could be virtual

- Develop short courses before/after conferences. Supported by NATO. Tied to codes, standards and legislation (at least partly) and adapted to national needs

- Problem statement from NATO that identifies education as being important for health

- Harmonization of legislation

- Identify a responsible party, although, the institution could be virtual

- Directive from NATO to countries

- East-West transfer of ideas and technology

- Training of the trainers

- Report at Indoor Air '99 International conference on the results of the workshop

- Quality assurance through existing specific international societies (e.g., International Society of Public Health - ISPH)

- Develop a volume plus a list-server (e.g., educational tool self-paced/self taught). Link to institution for quality assurance and incentives (depending on the level of education)

5.   Recommendations

The workshop participants defined a set of recommendations as follows:

The topics of the core curriculum need to be developed with proper enphasis on different specializations and with different formats and depth for diverse audiences, from graduate students to the general public; as well as to develop a framework for sustainable technology transfer. Therefore, it is recommended that this initial successful effort should continue to be supported in its development, at least to the point where the product is fully developed and the mechanisms for its application are identified.

A systematic approach is recommended for promotion of the core curriculum; its implementation and to complete what remains to be done to develop its topics with proper emphasis for different specializations and with different formats and depth for diverse audiences.

Further, it was recognized that the action plans identified needed to be placed in short and long-term feasible scenarios. A specific analysis of those scenarios was not conducted. However, short-term plans were made to implement a few of the key actions that emerged from the Workshop. It was recommended that in summer 1999 on the occasion of Indoor Air '99, the triennial Conference of the International Academy of Indoor Air Sciences, a workshop be held to: present the results of this project to a larger scientific community; broaden support for development and implementation; present the core curriculum and increase awareness.

Further, it was recommended that there be an additional workshop to refine the core curriculum (weighting of subjects) and develop a framework for technology transfer (To whom, what for and which institutional base) be organized in year 2000 as a follow-up to the Budapest '98 NATO workshop.

6.   Discussion and Conclusions

The workshop was very succefull in meeting the overall goal and objectives. A representative sample of existing educational programs currently available in the field of IAS at institutions of higher education, in government agency and other professional settings was discussed; current trends, approaches and emerging issues that are characterizing education in IAS were identified; players involved in IAS education were identified; educational needs in different cultural and geographical contexts were compared; a conceptual framework for international curricula was defined; possible strategies and action plans for establishing educational programs in IAS in different cultural and geographical contexts were identified.

The "forum session" that defined the tasks for the focus group sessions was much more contentious than the focus group sessions. The group interacted extensively to define the tasks. Although different viewpoints emerged, consensus was reached on proceeding to development of the core curriculum. Consensus was based on the recognition that the topics to be identified are those that everyone needs to know at least at some level.

The degree of consistency in the final lists of topics to be included in a core curriculum developed by two groups independently provides a sense of validity for the resulting topics. The relative ease with which the final product was agreed on is an indication that, at least at the core topical level, there is a remarkable amount of agreement across indoor air scientists from a number of fields, professions and countries.

Overall, the Workshop's conclusion was that a core curriculum for IAS was identified. How detailed each topic has to be developed and how much emphasis each would have relative to other topics has to be determined based on the audience, profession or objective of a course. Obstacles to its implementation exist at the policy, academic and professional level and the strategies identified need to be considered in short and long-term feasible scenarios. Plans were made to implement a few of the key recommendations emerged from the Workshop.

The Workshop offered an opportunity to bring together 28 participants coming from 14 different countries in a wide range of disciplines and professions. In doing so, the project originated, and reinforced existing patterns of communication that facilitate more effective transfer of science and technology between educators, scientists and policy makers. The Workshop provided a forum to conduct a critical assessment of existing educational paradigms in IAS and to define a conceptual framework for international curricula that could be integrated in existing educational programs. The Workshop promoted and underlined the importance of the interactions between educators in various fields of IAS and constituted a unique opportunity for identifying global issues of concern and future research agenda in IAS education. Further, it contributed toward the further development of a friendly international research/educational and professional cooperation by bringing about a better understanding of the principles that guide different cultural settings.

Finally, it should be noted that the conclusions and recommendations of this study are based on the knowledge and expertise available at this time. A great need for research and further understanding of educational needs in IAS still remains. It is my hope that this book will be a call to action. Many of the adverse health effects caused by indoor environments can be prevented and education has been proven to be an effective tool for prevention. What is needed is a plan that aims to raise public awareness and ensure appropriate education and training for professionals involved in the health and building sector.

AUTHOR INDEX

SUBJECT INDEX

252